T0301015

SHIFT REGISTER SEQUENCES

Secure and Limited-Access Code Generators,
Efficiency Code Generators, Prescribed Property
Generators, Mathematical Models

Third Revised Edition

SHIFT REGISTER SEQUENCES

Secure and Limited-Access Code Generators, Efficiency Code Generators, Prescribed Property Generators, Mathematical Models

Third Revised Edition

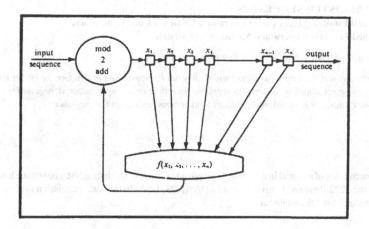

Solomon W. Golomb
University of Southern California, USA

World Scientific

NEW JERSEY · LONDON · SINGAPORE · BEIJING · SHANGHAI · HONG KONG · TAIPEI · CHENNAI · TOKYO

Published by

World Scientific Publishing Co. Pte. Ltd.
5 Toh Tuck Link, Singapore 596224
USA office: 27 Warren Street, Suite 401-402, Hackensack, NJ 07601
UK office: 57 Shelton Street, Covent Garden, London WC2H 9HE

British Library Cataloguing-in-Publication Data
A catalogue record for this book is available from the British Library.

SHIFT REGISTER SEQUENCES
Secure and Limited-Access Code Generators, Efficiency Code Generators,
Prescribed Property Generators, Mathematical Models

For photocopying of material in this volume, please pay a copying fee through the Copyright Clearance Center, Inc., 222 Rosewood Drive, Danvers, MA 01923, USA. In this case permission to photocopy is not required from the publisher.

ISBN 978-981-4632-00-3

Printed in Singapore

Solomon Golomb passed away in early May, 2016,
only days after completing the revisions for this new
edition. His wife Bo died two weeks later. We — their family,
friends, colleagues and publisher — dedicate this book
to the memory of a wonderful couple, with
admiration, gratitude and love.

INTRODUCTION TO SOLOMON W. GOLOMB

E. Berlekamp

University of California, Berkeley

berlek@math.berkeley.edu

I feel greatly honored to write this Introduction to the Proceedings of the Golomb70-fest.[1] Prof. Golomb is well known as the recipient of the highest honors possible at University of Southern California, the Information Theory Society, and the US engineering profession.[2]

Solomon W. Golomb was born on May 30, 1932.[3] His father was a rabbi and a linguist. Sol soon developed a precocious appetite for mathematics as well as for languages and a wide range of classical literary works. Sol completed his undergraduate studies at Johns Hopkins in two years, then obtained a PhD in mathematics from Harvard, and went to Norway on a Fulbright fellowship, where he met his future bride.[4] Along the way, he spent a summer working for Martin Aircraft Company, where he also became interested in a variety of engineering problems in aerospace electronics. In pursuit of this interest, he moved to Southern California to begin his first full-time job with the Jet Propulsion Laboratory. This was in the early heyday of NASA, immediately following sputnik. Sol soon emerged as the leader of JPL's Space Communications efforts.

My first encounter with Golomb's name was as the author of a paper published in the proceedings of a 1962 symposium, entitled, "Mathematical Theory of Discrete Classification." I thought this was a brilliant paper, and it had a big impact on me. I was surprised and flattered when I was invited to meet Sol at JPL in January 1965, and only thereafter did I come to realize the wide breadth of his work, which included many other papers and books that were generally regarded as even more significant than the one that had so strongly impressed me. By January 1965, Sol had already moved from JPL to his professorship in both mathematics and electrical engineering at the University of Southern California, but he remained an influential eminence grise at JPL. Largely on his recommendation, JPL hired me as a consultant. I visited there weekly for the next year and a half, and I was able to work personally with an extraordinary cast of characters, many of whose names appear in this volume. The two who worked most closely with me, Gus Solomon and Ed Posner, are now both deceased. There was also a bright array of luminary consultants, including Lloyd Welch, Andy Viterbi, Irwin Jacobs, Marshall Hall, and some Caltech students, including Bob McEliece and Richard Stanley. Sol was heavily involved in the recruitment of most of these people, as well as many others. On the occasion of Sol's 60th birthdayfest in 1992, Gus Solomon proposed a memorable toast to "the man who brought modern combinatorial mathematics to Southern California." At the time, I was shocked by the boldness of the claim. After some further reflection, I was also shocked by the surprisingly large amount of truth it contained.

Sol has written landmark books on a wide variety of topics. "Shift Register Sequences" are used in radar, space communications, cryptography and now cell phone communications. This book has long been a standard reading requirement for new recruits in many organizations, including the National Security Agency and a variety of companies that design anti-jam military communication systems. "Polyominoes" defined that subject and established Sol as a leader in the broad field of recreational mathematics. Both "shift register sequences" and "polyominoes" have become subject headings in the classification of mathematics used by Mathematical Reviews. "Digital Communications with Space Applications," written with Baumert, Easterling, Stiffler, and Viterbi, was among the earliest and most influential books on that subject. "Information Theory and Coding," written with Peile and Scholtz, is a novel text for graduate course which has become very popular at USC and elsewhere.

Sol maintains a strong interest in elementary number theory. Many of his research papers deal with questions concerning prime numbers. Sol also has a great interest in teaching, both advanced and elementary courses, and in promoting popular interest in mathematics. He is an avid collector, solver, and composer of problems. He has authored the Problems Section (sometimes known as Golomb's corner) in periodicals including the Newsletter of the IEEE Information Theory Society, the Alumni Magazine of Johns Hopkins University, and the Los Angeles Times. Following a change of publishers some years ago, Sol's column was discontinued. But, he noted, this was totally consistent with the new editorial policy then being adopted, which also discontinued the entire Science News section and increased the coverage of astrology.

Sol's publications also include some provocative commentaries on the philosophy and history of mathematics. His 1998 critique of G. H. Hardy's famous 1940 "Mathematician's Apology" ("Mathematics Forty

Years After Sputnik") is a well-documented and articulate exposition of the practical values of "pure" mathematics, including even those topics discovered by purists like Hardy. In some respects, Sol's 1982 obituary of Max Delbrook may be a more objective view of the rise of modern molecular biology than could have been written by any of the major pioneers in that field. Golomb's interest in biology predates even his own paper, "On the plausibility of the RNA code," published in the 1962 issue of Nature, long before the idea of a lengthy digital basis of inheritance had appeared on the mental radar screens of most biologists.

Sol has also been a senior academic statesman. He served as President of USC's Faculty Senate in 1976-1977, as Vice Provost for Research in 1986-1989, and as Director of Technology at USC's Annenberg Center for Communications in 1995-1998. He founded the "National Academies Group" at USC, and restored the influence of the faculty on the governance of that university on certain occasions when the administrations of the day had been veering off course.

I have personally had the great fortune to be placed in positions which allowed me to see certain other aspects of Sol's multi-faceted intellect. Annoyed by the commercial misappropriation of one of his polyominoe games, in the late 1960s, Sol incorporated one of his hobbies into a company called "Recreational Technology, Inc." I accepted his invitation to become a founding director of this venture. Although this venture never had any outside investors, nor any significant sales or earnings, it became the dry run for another venture called Cyclotomics, which, at Sol's urging, I founded in December 1973. Sol was my founding outside director and primary business mentor and confidante for the next 15 years.

Notes

1. This Introduction from 2003 is reprinted here from *Mathematical Properties of Sequences and Other Combinatorial Structures*, edited by Jong-Seon No, Hong-Yeop Song, Tor Helleseth, P. Vijay Kumar, in The Springer International Series in Engineering and Computer Science, Volume 726 (2003) with permission of Springer.
2. These honors include, among others, USC's Presidential Medallion, memberships in the National Academy of Science and of Engineering, the IEEE Hamming Medal and Shannon Award, the Benjamin Franklin Medal, and the National Medal of Science — the country's highest distinction for contributions to scientific research.
3. May 31, 1932 was his actual birth date.
4. He met his future wife "Bo" on a side trip to Denmark.

References

[1] S. W. Golomb, "Mathematical Theory of Discrete Classification", in *Information Theory*, Proceedings of the Fourth (1960) London Symposium, Colin Cherry, Editor, Butterworths, London, 1961.

[2] S. W. Golomb, "Mathematics Forty Years After Sputnik", *American Scholar*, vol. 67, no. 2, Spring, 1998.

[3] S. W. Golomb, "Mathematics After Forty Years of the Space Age", Mathematical Intelligencer, vol. 21, no. 4, Fall, 1999.

[4] S. W. Golomb, "Max Delbruck - An Appreciation", American Scholar, vol. 51, no. 3, Summer, 1982.

PREFACE

The theory of shift register sequences has found major applications in a wide variety of technological situations, including secure, reliable and efficient communications, digital ranging and tracking systems, deterministic simulation of random processes, and computer sequencing and timing schemes. Yet this theory has been presented previously only in disjointed and scattered form, in a variety of out-of-print or otherwise inaccessible company reports, and in scattered journal articles. The purpose of this book is to collect and present in a single volume a thorough treatment of both the linear and nonlinear theory, with a guide to the area of application, and a full bibliography of the related literature.

From an engineering viewpoint, the theory of shift register sequences is very well worked out and fully ready for use. However, from a mathematical standpoint, there are certainly many unresolved problems worthy of further study.

I was first introduced to the problem of shift register sequences in 1954, while on a summer job with the Glenn L. Martin Company in Baltimore; and that was the beginning of a long friendship. My initial reaction was that the mathematics involved was extremely beautiful and that unfortunately the application was probably just shortlived and insignificant. Little did I realize then how important shift register techniques were destined to become in our technology.

I returned to Harvard in the fall, and I took advantage of being in Cambridge to pay frequent visits to Neal Zierler and others at the Lincoln Laboratory of M.I.T. who were similarly interested in shift registers. Finally, in June, 1955, I submitted my paper "Sequences with Randomness Properties," as the final progress report on my consulting contract with the Martin Co.

In the summer of 1956, I took a position at the Jet Propulsion Laboratory, where there was already considerable interest in shift register sequences. At JPL I worked both individually and in collaboration with Lloyd Welch on these problems for several years, and a number of important reports were produced. The transfer of the JPL contract, in 1958, from Army Ordnance to NASA, meant a major redirection in the type of *applications* we sought for shift register sequences, but, as it turned out, the theoretical foundation was already almost complete. One of the many major applications of this work has been to a remarkably precise interplanetary-distance ranging system, which has been adopted by the Deep Space Network, operated by JPL for the Office of Tracking and Data Acquisition of NASA.

It is hard to establish accurate priorities as to who did what first. For example, E. N. Gilbert of the Bell Telephone Laboratories derived much of the linear theory a year or so earlier than either Zierler, Welch, or myself, but his memorandum had very limited distribution. Many others have rederived the linear theory independently since that time, and doubtless others will continue to do so. Of course, the first investigatio.: of linear recurrence relations modulo p goes back as far as Lagrange, in the eighteenth century, and an excellent modern treatment was given (as a purely mathematical exposition) by Marshall Hall in 1937.

In assembling this volume, the procedure which I followed was to take the most important reports and articles of which I was an author or co-author, and edit them into a systematic exposition, with a reasonable continuity of both style and subject matter.

To preserve the historical flavor and continuity of the material, I have indicated the original publications from which the various papers were extracted, and in those relatively rare cases where the original version contained errors of fact or flaws in reasoning, I have not hesitated to correct them.

The two chapters which form Part One are chronologically the most recent. However, they are written from a tutorial standpoint, and thus serve to put the more technical material which follows into better perspective. Also, being more recently written, they give a more up-to-date indication of the place of shift register theory and applications in our current technology. Chapter II indicates the broader framework (namely, mathematical machine theory) within which shift registers are such an important special case.

Part Two deals with the linear theory. Although the term *linear* has been much overworked and abused, there is a reasonably consistent usage common to the terms *linear algebra, linear differential equations, linear operators, linear difference equations,* and *linear systems theory.* Moreover, there are a number of standard techniques for analyzing linear systems–matrix methods, operator methods, Laplace-Stieltjes transform methods, flow graph methods, impulse-response methods, etc.

All of these methods succeed in replacing a linear system by its "characteristic equation," and the behavior of the system is related to the factorization and the roots of the characteristic equation.

Linear shift registers are *linear* in this standardized sense. However, the underlying arithmetic is not that of the real or complex numbers, but of the field of two elements, 0 and 1, operating modulo 2. The analysis of linear shift register behavior, then, reduces to the study of their characteristic equations, which are polynomials with coefficients in the field of two elements.

In Chapter III, "Sequences with Randomness Properties," we start with certain desirable constraints on binary sequences, and are led to linear shift registers for the generation of such sequences. The analysis of linear shift registers uses the method recently popularized among electrical engineers under the name of the "Z-transform," and known to mathematicians since the late eighteenth century as the method of "generating functions." It would have been equally valid to use any of the other methods for analyzing linear systems, and there are articles by various authors which do so. In any case, the same theorems result, and the same correspondence between shift registers and polynomials occurs.

In Chapter IV, "Structural Properties of PN Sequences," we pay special attention to the correlation properties and spectra of shift register sequences. The correlation results provide a deeper insight into the behavior of the linear shift register sequences, and the spectral results are particularly important when the shift register sequence is used to modulate a radio signal.

Finally, in Chapter V, "Factorization of Trinomials over GF(2)," there is a detailed discussion of the theory of polynomial factorization over the field of two elements, with special emphasis on trinomials (i.e. three term polynomials), which correspond to the simplest shift registers to construct. Chapter V concludes with a factorization table for trinomials through degree 46.

Part Three deals with the nonlinear theory. The "nonlinear" case is, of course, the general case, in which almost anything can happen. Unlike the highly restricted "linear" case, where we developed the necessary analytical procedures to determine exact lengths of sequences, it is sufficiently ambitious in general to ask *qualitative* questions.

In Chapter VI, "Nonlinear Shift Register Sequences," we concern ourselves with such problems as when the state diagram has only "pure" cycles, without branches, and how often all the states lie on a single cycle. In the case of branchless cycles, a simple criterion is given for whether the total number of cycles is even or odd, and a number of important corollaries are deduced from this result.

In Chapter VII, "Cycles from Nonlinear Shift Registers," we develop a *statistical* model for the number of cycles, the expected lengths of the cycles, the probability that a given vector lies on the longest cycle, etc. Also, a construction is explained for obtaining cycles of any length from 1 to 2^n inclusive from a shift register of n stages.

A nonlinear shift register necessarily involves a Boolean function of n variables to compute the feedback term. We conclude the discussion of the nonlinear case with Chapter VIII, "On the Classification of Boolean Functions," which explains the reduction in the number of truly distinct cases which need to be considered, based on symmetry properties of the Boolean functions.

Not surprisingly, the nonlinear theory leaves many important questions as yet unanswered. However, the treatment presented here resolves most of the basic qualitative issues, and sets up procedural guidelines and methodology for further investigations.

PREFACE
TO THE REVISED EDITION

In the fifteen years since *Shift Register Sequences* was originally published in hard cover by Holden-Day, Inc., a great many developments have taken place. Far more is now known about both the theory and the applications of these sequences, and it would be a monumental undertaking to revise the book so as to reflect all of this. The present objective is more modest. There has been great demand for copies of the book since it went out of print, and even for the long-unobtainable Martin Co. and Jet Propulsion Laboratory reports which were reincarnated as some of its chapters. To fill this demand, the paperback edition faithfully reprints the original text, with two significant additions. One is a Comprehensive Bibliography with more than 400 entries, which gives the ambitious reader a head start on getting fully up to date in those areas in which he is most interested. The other is a new Chapter 9, titled *Selective Update*, which describes some of the most important recent developments involving topics which are already treated in the text. It is my hope to publish another volume, based on nine or ten of the most important papers which have appeared about shift register sequences since 1967, within the next two years.

I wish to express my gratitude to Wayne G. Barker, President of the Aegean Park Press, for his interest in publishing this edition, and for prodding me into doing the required work; and to Holden-Day, Inc., and its President, Fredrick H. Murphy, for the assignment of copyright from the original edition.

Solomon W. Golomb
Los Angeles, California
May 1, 1982

PREFACE TO THE THIRD EDITION

Since 1982, when the Second Edition of my *Shift Register Sequences* book
was published by (now defunct) Aegean Park Press, the amount of new
related material is so vast that it could fill an entire bookcase. Accordingly, in preparing this new edition, I have retained the first eight chapters
(which were already in the first, 1957, edition, and have stood the test
of time), but I have extensively revised Chapter IX, titled SELECTIVE
UPDATE, from the form in which it appeared in the Second Edition. I
have mentioned quite a few more recent developments, but I have generally
tried to stay within the general subject areas of the previous editions. It
is now sixty years since my first widely circulated article on this subject
appeared ("Sequences With Randomness Properties," a 1955 report from
Glenn L. Martin Co., now part of Lockhead-Martin) which is the basis of
Chapter 3 of this book and its previous editions. This has served well as an
introduction to shift register sequences to several generations of engineers,
mathematicians, crytographers, and computer scientists, and I hope it will
continue to do so far many years to come.

Solomon W. Golomb

Los Angeles, California, USA
May, 2016

ACKNOWLEDGMENTS

I would like to express my gratitude and appreciation to both the Martin Company and the Jet Propulsion Laboratory for permission to reprint material which originally appeared in company reports, as follows:

"Sequences with Randomness Properties," Martin Co., June, 1955;
"Nonlinear Shift Register Sequences," JPL, October, 1957;
"Structural Properties of PN Sequences," JPL, March, 1958;
"On the Factorization of Trinomials Over GF(2)," JPL, July, 1959;
"Cycles from Nonlinear Shift Registers," JPL, August, 1959.

Other material which I have incorporated includes two papers which I had presented at technical meetings, "On the Classification of Boolean Functions." from the Transactions of the International Symposium on Circuit and Information Theory, IRE Transactions on Circuit Theory, May 1959, and "The Present Status of the Shift-Register Art," from the URSI meeting in Washington, D.C., in May 1962. The latter was subsequently incorporated into Chapter I, Section 5, of the book *Digital Communications* (Golomb et al., Prentice-Hall, 1964), and permission by the publisher to reprint this material is gratefully acknowledged. Finally, the material on "The Shift Register as a Finite State Machine" is based on material contained in Technical Report 122, University of Southern California, School of Engineering, December, 1964.

A special acknowledgment is due to Dr. Lloyd R. Welch, Associate Professor of Electrical Engineering at the University of Southern California, who was co-author of the three JPL reports which appear here as Chapters 5, 6, and 7. We further had the collaboration of Dr. Alfred W. Hales, Assistant Professor of Mathematics at UCLA, in the writing of Chapter 5 and of Dr. Richard M. Goldstein, Chief of the Communications Systems Research Section at JPL, in the writing of Chapter 7. (Needless to say, we all had less exalted titles at the time the reports were originally written.) I must also thank Mr. Harold Fredricksen for assistance at many points along the way, from generating data and tables in the original reports down to correcting errors in the galley proof for this book.

I am also grateful to Holden-Day for the encouragement to embark upon this venture.

S. W. G.

University of Southern California
Los Angeles

CONTENTS

PART ONE

PERSPECTIVE

Chapter I

THE PRESENT STATUS OF THE SHIFT REGISTER ART

1. INTRODUCTION

For nearly two decades, the idea of using shift registers to generate sequences of 1's and 0's has been explored, developed, and refined. At the present time, a wide variety of basically different applications of shift registers are either under consideration or in actual use. It is the purpose of this chapter to list and briefly to describe the principal areas of application. In addition, it will indicate some of the properties of shift register sequences which make them well-suited to these various applications.

2. AREAS OF APPLICATION

2.1. *Secure and limited-access code generators*

Encipherment: It has been suggested that a message written in binary digits could be added, modulo 2, to a shift register sequence acting as the "key." The decipherment consists of adding the key modulo 2 to the coded message. This type of cipher (called the Vigenère cipher) is only one of many classical encipherments in which a shift register sequence could play the role of the key.

Privacy encoding: There are many situations where one wishes to communicate a message (most typically a command) through a hostile environment, where the information need not remain secret after it is received and has been acted upon. The use of shift register sequences in these situations to provide positive authentication of the message is an attractive possibility.

Multiple address coding: Different portions of a long shift register sequence can be assigned as characteristic *addresses* to a large number of individuals, aircraft, substations, and so forth. These addresses may be used either for positive identification, or merely to enable a large number of scattered messengers to report systematically to their

2

home base. For example, such a system, assigning ten-bit segments of a sequence of length 1023 to each of 64 outlying weather stations, was installed to monitor rainfall information in the vicinity of Calcutta [25].

2.2. *Efficiency code generators*

Error-correcting codes: Many and varied schemes for error-correcting codes based on linear shift register sequences have been proposed and studied. In conventional coding terminology, these may all be classified as *cyclic, parity-check, group codes*. An extensive discussion of this subject, including the use of shift registers to implement Bose-Chaudhuri codes, is contained in W. W. Peterson's monograph [29].

Signals recoverable through noise: For range radar applications in environments with extreme background noise, a shift register modulated pulse-train or CW signal, using a maximum length sequence, has the property that its autocorrelation function is recoverable despite a noise-to-signal excess of many db. (The exact figure depends on the integration time available in the correlator.) For example, the early Venus ranging experiments of Lincoln Laboratory used a maximum length linear sequence of length $2^{13} - 1 = 8191$ to determine whether to transmit "pulse" or "no pulse" in consecutive time intervals. In contrast, the JPL ranging system uses a Boolean combination of several short sequences, which facilitates rapid acquisition, and the combination is used to specify biphase modulation on a CW carrier.

2.3. *Prescribed property generators*

Prescribed period generators: It has recently been proved that with a shift register of degree n, all sequence lengths p between 1 and 2^n (inclusive) can be obtained by use of a suitably chosen feedback logic. A simple procedure for finding and constructing this logic has also been described. A typical application is to the construction of a countdown circuit within a synchronous digital system.

Prescribed sequence generators: It is reasonable to explore the possibilities of using a shift register as a memory device, so that its output is some preassigned sequence. This can be attempted either *internally* (where the feedback logic is so specified that the sequence actually running through the shift register is the prescribed sequence), or *externally* (where the feedback logic is used to generate a sequence of the desired period, as in the preceding paragraph, and an external logical function is used to transform the sequence of desired period into the preassigned sequence). With either approach, it is necessary to finish specifying a partially filled-in truth table in such a way as to minimize the complexity of the resulting Boolean Interpolation Problem. A number of computational routines for solving the Boolean Interpolation Problem exist, and some of these methods use to special advantage the fact that the problem involves a shift register sequence. Thus it is possible to *store* data of many different sorts in shift register form.

2.4. *Mathematical models*

Random bit generators: A shift register serves as an approximate model for the outcome of a coin-tossing experiment, despite the underlying determinacy of the system. In fact, axiomatizations of statistics based on sequences of length 2^n in which every n-bit combination occurs exactly once (i.e., de Bruijn's shift register sequences) have been seriously proposed. Moreover, the use of shift registers to generate random numbers for Monte Carlo statistical techniques has been advocated by several mathematicians.

Finite state machines: The shift register with feedback serves as the simplest non-trivial example of a finite state machine without external inputs. The shift register thus serves as an elementary "experimental animal" for research in recursive logic machines and in automata studies.

Markov processes: The de Bruijn diagram for the possible sequential behavior of an n-stage shift register is also the Markov state-diagram for a binary channel with statistical dependence limited to n bits into the past. The shift register usage is a limiting case of the Markov process in which only the transition probabilities 1 and 0 (pure determinism) are allowed. However, topological or graph-theoretic properties of the diagram discovered by shift register research is interpretable in terms of the most general Markov process situation.

3. ADVANTAGES OF SHIFT REGISTERS

3.1. *Secure code generators*

Detailed discussion of this point would exceed the scope of an unclassified exposition. The reader interested in an excellent unclassified discussion of the general question of secure coding from the viewpoint of information theory should see Shannon's paper [35].

3.2. *Efficiency code generators*

Maximum length linear shift register sequences have a vast amount of internal combinatorial structure. At the heart of this structure is an orthogonality constraint between the various time-shifts of the sequence. For error-correcting codes, this constraint leads to the possibility of linearly independent parity checks. For signals recoverable through noise, it leads to perfect two-level auto-correlation, the ideal shape for this purpose. For both kinds of applications, the shift register sequences have the further important advantage of being remarkably simple to generate.

3.3. *Prescribed property generators*

The first important property of shift registers for these applications is their generality, in the sense that every finite (or periodic) binary sequence is obtainable from some suitably constructed shift register. Next, not only an analysis for describing the sequence given the shift register configuration, but also a synthesis for specifying the configuration given the sequence, is available.

There is considerable storage efficiency associated with certain of these shift register applications. The use of the prescribed period generator as a countdown circuit is the most efficient mechanization possible, in a rather precise sense. Shift register memory could be preferable to other types of memory devices in systems constructed primarily from digital building blocks, since the shift register could be assembled from the same kind of blocks, and the tie-in to the system would be direct, without the problems of access, synchronization, compatibility with voltage and impedance levels, and so forth.

In comparison with other digital generation schemes, for example, that use a counter to generate a prescribed period, and an external logic to transform this into a preassigned sequence, the shift register approach has several advantages. First, from a logical machines standpoint, a shift register is a simpler and more natural device than a counter. This is reflected in the simplicity of transition from state to state, without delays required for carry digits to propagate. Second, the different states in the shift register are merely delays of one another, so that in constructing an absolute minimized logical function external to a shift register, any delays required by this function are already available from the shift register. In the case of a counter, on the other hand, extra delays would have to be built into the external logical function. Finally, to the extent that one can accomplish the desired objectives with the simplest type of finite state machine (i.e., a shift register), it is not necessary to use more involved mechanizations.

3.4. *Mathematical models*

Any deterministic device for generating allegedly random bits is open to obvious criticism. On the other hand, no finite sequence is ever truly *random*. ("Randomness" in fact properly refers, in the finite case, to the selection procedure whereby the choice was made, rather than to the chosen object itself.)   One may instead define *symptoms* or *criteria* of randomness, and accept as pseudo-random any sequence passing the corresponding randomness tests.   Linear shift register sequences which are quite easy to generate will pass the tests for run properties and for normality properties up to span n (i.e., each possible block of m bits occurring with approximate probability 2^{-m} for $m < n$).   For most applications of Monte Carlo techniques this is adequately random— certainly more effective than the technique of squaring a ten-bit number and keeping the middle ten bits.   In some applications, there are also advantages to having a "standard" pseudo-random sequence, to be used repeatedly, so that different systems may be subjected to the same "noise environment."

The shift register as a typical finite-state machine has also been studied extensively, and many of the most important properties of shift registers (e.g., when all the cycles are "pure" or "branchless," or in coding terminology when the device is "information-lossless") have been generalized to a far broader class of logical machines.   This subject is treated further in Chapter II of this book.

The shift register as a deterministic limiting case of the binary Markov process has not yet been fully exploited.   The theorem that an n-stage shift register, with suitable feedback logic, is capable of producing any period p, $1 \leq p \leq 2^n$, is really a more general theorem about binary Markov processes: namely, that if all transitions are possible, there exists a simply connected loop for every p, $1 \leq p \leq 2^n$, which returns to the starting point after exactly p transitions.   The de Bruijn diagram for possible shift register sequences is homeomorphic (topologically equivalent) to the binary Markov diagram.   Since the "prescribed period theorem" has been proved by shift register methods, rather than by purely topological means, the potential utility of using shift register theory to study Markov processes is evident.

Chapter II

THE SHIFT REGISTER AS A FINITE STATE MACHINE

1. FINITE STATE MACHINES

1.1. *Definition*

The mathematical notion of the *finite state machine* encompasses most special purpose digital computers, calculating machines, encoders and decoders, and so forth. By definition such a device consists of a finite collection of *states* $K = \{K_i\}$ and sequentially accepts a sequence of *inputs* from a finite set $A = \{a_i\}$, and produces a sequence of *outputs* from a finite set $B = \{b_i\}$. Moreover, there is an *output function* μ which computes the present output as a fixed function of present input and present state, and a *next-state function* δ which computes the next state as a fixed function of present input and present state. Thus $\mu(K_n, a_n) = b_n$, while $\delta(K_n, a_n) = K_{n+1}$. In abstract notation, μ is a function from $K \times A$ into B, and δ is a function from $K \times A$ into K.

1.2. *Two representations of finite state machines*

A finite state machine is completely specified by a *function table*, which indicates the output and next state corresponding to every possible combination of state and input. An example of such a table is shown in Figure II-1, where the corresponding machine has possible states R, S, and T, accepts the input symbols 0 and 1, and produces the output symbols α, β, and γ.

If we are told the initial state of a machine, and are given a specified input sequence, we can compute the output sequence and determine the terminal state of the machine (i.e., the state of the machine after processing the last input symbol). For example, if the machine specified in Figure II-1 has the initial state R and is fed the input tape 01100101, we can keep track of what happens as shown in Figure II-2. Here we see that the output sequence is $\alpha\gamma\gamma\beta\alpha\gamma\beta\beta$ and the terminal state is T.

An alternative to Figure II-1 for representing finite state machines

	0	1
R	S, α	T, β
S	R, β	S, γ
T	T, α	R, α

Fig. II-1. The function table of a finite state machine.

Time	Input	State	Output
1	0	R	α
2	1	S	γ
3	1	S	γ
4	0	S	β
5	0	R	α
6	1	S	γ
7	0	S	β
8	1	R	β
9		T	

Fig. II-2. Sequential behavior of a finite state machine.

is the *state diagram* method shown in Figure II-3. Here the three states of the machine are shown as nodes, and the input symbols produce oriented paths from node to node in accordance with the next-state function. On each edge of the graph we write in parentheses the output symbol corresponding to that particular symbol. Thus, if we are in state *R* and receive the input symbol 0, the diagram shows that we progress to state *S*, and write the output symbol α; but if we are in state *R* and receive the input symbol 1, we proceed to state *T*, and write the output symbol β.

In principle, a finite state machine is completely specified by either its function table or its state diagram. The function table is a tidier

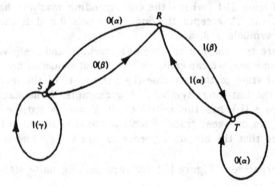

Fig. II-3. The state diagram of a finite state machine.

method of storing the description of the machine, but the state diagram is easier to follow when tracing out the computations of the machine on a given input sequence.

1.3. *Equivalence of two machines*

From a practical standpoint, it is important to know when two different machines will perform exactly the same work. Given such a notion of *equivalence* of two machines, we can then ask which of all the machines equivalent to a given one is *simplest* (i.e., has the smallest number of states).

Two machines which accept the same set of input symbols and write the same set of output symbols are called *equivalent* if and only if both machines perform the same transformation from the set of input sequences to the set of output sequences. That is, they are equivalent if and only if they "reduce" or "process" data so as to yield identical results.

In Figure II-4, we see a state diagram which is more complicated than the one in Figure II-3 and involves four states instead of three. However, it is quite easy to verify that the machines of Figures II-3 and II-4 are equivalent. Specifically, if we regard T and U as being merely two names for the same state, we observe that Figure II-4 condenses to Figure II-3, without in any way affecting the correspondence between input sequences and output sequences.

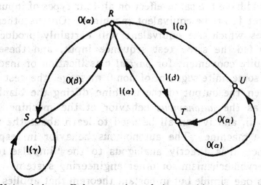

Fig. II-4. A less efficient version of the machine in Figure II-3.

In general, several algorithms are known which can determine whether or not two finite state machines are equivalent, and which can help to find, among all machines equivalent to a given machine, the one with the fewest states. For further information concerning these problems, the reader is referred to the bibliography, particularly references [12] and [24].

1.4. *Analysis and autonomous behavior of machines*

The *behavior* of a finite state machine can only be discussed with reference to the processing which it performs on various input sequences. Since even the number of possible *finite-length* input sequences is infinite (and the number of infinite-length input sequences is non-denumerably infinite), it is clearly impractical to "test" the machine's action on *all possible* inputs. However, it is quite reasonable and useful to test the machine's effect on certain extremely simple input sequences. Some of the natural ones to try are the following:

    i. The "blank" tape, consisting of an infinite input sequence of 0's.

    ii. A "constant" tape, consisting of an infinite input sequence of 1's (or some other non-0 value, if there are more than two possible input symbols).

    iii. The "impulse" tape, consisting of the symbol 1 followed by an infinite sequence of 0's.

    iv. The "step input" tape, consisting of a finite number of *ones* (or some other constant input value), followed by an infinite sequence of zeros.

The interaction of the machine with the types of inputs just listed is not sufficient to characterize the machine completely. Thus, two machines could have the same effect on all four types of input sequences listed, and yet fail to be equivalent machines. On the other hand, any two machines which are equivalent will certainly produce the same output when fed the same test sequence input, and these particular inputs are quite convenient for *partial* classification of machines.

If, after some finite segment of non-0 message, the rest of the tape is blank, then the output of the machine during the blank portion is referred to as the *autonomous* behavior of the machine. The input sequences i, iii, and iv can all be used to learn about the autonomous behavior of a machine. The autonomous behavior in response to the input sequence iii is directly analogous to the "impulse response" of a filter, a servomechanism, or other engineering system.

There is one simple but important theorem that applies to all finite state machines and all four of the special input sequences listed, viz., that the output sequence will ultimately become (and remain) periodic. The possible periods of the output sequences so generated are important *invariants* in the analysis and classification of finite state machines. Formally, we have:

Theorem 1. If the input to a finite state machine is constant (except perhaps for a finite initial segment), then the output is ultimately periodic.

Proof. There are only finitely many states. Hence, during the constant portion of the input sequence, the machine will sooner or later return to a previous state. If the machine is in state S at both times t_1 and t_2, it will be in the same state S' at both times $t_1 + 1$ and $t_2 + 1$ (since the same input symbol and same state generates the same *next* state), and in this fashion a periodicity is established.

More generally, the same argument can be used to prove:

> **Theorem 2.** If the input sequence to a finite state machine is ultimately periodic, then the output sequence is ultimately periodic.

It is important to realize, however, that the periods of the input and output sequences are likely to differ a great deal.

1.5. *Examples*

Let us take the machine of Figures II-2 and II-3 in the initial state R, and test its action upon the four special types of input sequences listed in the previous section.

i. If $000000\ldots$ is the input sequence, then $\alpha\beta\alpha\beta\alpha\beta\alpha\beta\ldots$ is the output sequence, and $RSRSRS\ldots$ is the sequence of states.

ii. If $111111\ldots$ is the input sequence, then $\beta\alpha\beta\alpha\beta\alpha\beta\alpha\ldots$ is the output sequence, and $RTRTRT\ldots$ is the sequence of states.

iii. If $100000\ldots$ is the input sequence, then $\beta\alpha\alpha\alpha\alpha\alpha\ldots$ is the output sequence, and $RTTTTT\ldots$ is the sequence of states.

iv. If $\underset{n}{\underbrace{11\ldots1}}000\ldots$ is the input sequence, there are two basically different cases. If n is odd, the output sequence is $\underset{n}{\underbrace{\beta\alpha\beta\alpha\ldots\beta}}\alpha\alpha\ldots$ where the sequence of states is $\underset{n}{\underbrace{RTRT\ldots RT}}TT\ldots$; while if n is even, the output sequence is $\underset{n}{\underbrace{\beta\alpha\beta\alpha\ldots\beta\alpha}}\alpha\beta\alpha\beta\alpha\beta\ldots$, where the sequence of states is $\underset{n}{\underbrace{RTRT\ldots RT}}RSRSRS\ldots$.

In Figure II-5, we see part of the state diagram of a decimal division machine, specifically the part that responds to the command: "divide by 7." To divide n by 7, we start in state n, and use the input sequence $77777\ldots$. Following the diagram, we can readily check the following computations:

$0/7 = 0\,0\,0\,0\,0\,0\,0\,0\,0\,0\ldots$

$1/7 = 1\,4\,2\,8\,5\,7\,1\,4\,2\,8\,5\,7\ldots$

$$2/7 = 2\ 8\ 5\ 7\ 1\ 4\ 2\ 8\ 5\ 7\ 1\ 4\ \ldots$$
$$3/7 = 4\ 2\ 8\ 5\ 7\ 1\ 4\ 2\ 8\ 5\ 7\ 1\ \ldots$$
$$4/7 = 5\ 7\ 1\ 4\ 2\ 8\ 5\ 7\ 1\ 4\ 2\ 8\ \ldots$$
$$5/7 = 7\ 1\ 4\ 2\ 8\ 5\ 7\ 1\ 4\ 2\ 8\ 5\ \ldots$$
$$6/7 = 8\ 5\ 7\ 1\ 4\ 2\ 8\ 5\ 7\ 1\ 4\ 2\ \ldots$$
$$7/7 = 1\ 0\ 0\ 0\ 0\ 0\ 0\ 0\ 0\ 0\ 0\ 0\ \ldots$$
$$8/7 = 1\ 1\ 4\ 2\ 8\ 5\ 7\ 1\ 4\ 2\ 8\ 5\ 7\ \ldots$$
$$9/7 = 1\ 2\ 8\ 5\ 7\ 1\ 4\ 2\ 8\ 5\ 7\ 1\ 4\ \ldots$$

The diagram clearly reveals that the possible periodicities for these computations are 1 and 6. Extending Figure II-5 to cover all divisions from 1 to 9 inclusive would lead to an extremely complicated graph to draw. However, it would not be difficult to portray the machine in function table form.

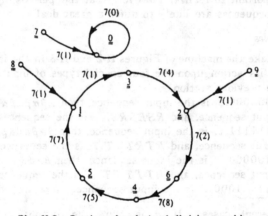

Fig. II-5.   Portion of a decimal division machine.

2.   SHIFT REGISTERS

2.1.   *Linear and nonlinear shift registers*

In Figure II-6 we see a diagram of a general (nonlinear) shift register with feedback. Each of the squares labeled $x_1, x_2, \ldots, x_n$ is a binary storage element (flip-flop, position on a delay line, or other memory device). At periodic intervals determined by a master clock, the contents of x_i is transferred into x_{i+1}. However, to obtain a new value for location x_1, we compute some function $f(x_1, x_2, \ldots, x_n)$ of all the present terms in the shift register and use this in x_1. (A function with n binary inputs and one binary output is called a *Boolean function of n variables*, and there are 2^{2^n} *different* Boolean functions for a given number n of variables.)

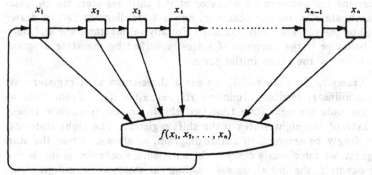

Fig. II-6. General diagram of a feedback shift register.

The n binary storage elements are called the *stages* of the shift register, and their contents (regarded as either a binary number or binary vector, n bits in length) is called the *state* of the shift register. At every clock pulse, there is a transition from one state to the next. Clearly, there are 2^n possible states (the binary numbers from $00\ldots0$ to $11\ldots1$) for a shift register of n stages. If we start the shift register in any one of these states, it then progresses through some sequence of states. As we shall see, this progression is really the autonomous behavior of a sequential machine; hence a periodic succession ultimately results, by Theorem 1 of the preceding section.

If the "feedback function" $f(x_1, x_2, \ldots, x_n)$ can be expressed in the form

$$f(x_1, x_2, \ldots x_n) = c_1 x_1 \oplus c_2 x_2 \oplus c_3 x_3 \oplus \ldots \oplus c_n x_n$$

where each of the constants c_i is either 0 or 1, and where the symbol $\oplus$ denotes addition modulo 2 (that is, 1 for odd sums and 0 for even sums), the shift register of Figure II-6 is called *linear*. Since there are 2^n ways to pick the n binary constants c_i, only 2^n out of the 2^{2^n} shift registers with n stages are linear. The properties of linear shift registers are developed extensively in Part Two of this book, and the nonlinear theory is presented in Part Three. For the present, however, we are principally concerned with the relationship of shift registers to the more general (and less explicit) theory of finite state machines.

2.2. *Shift registers as autonomous machines*

It is quite easy to describe the general shift register (Figure II-6) in terms of the theory of finite machines sketched in Section 1. An n-stage shift register is a finite machine with 2^n possible states. To determine the autonomous behavior of the shift register, the *successor* of each state is readily observed, and a state diagram can be drawn based on the successor information. For any initial state, the autonomous behavior is the sequence of states, specified by the state diagram, which follow the given initial state.

Example. In Figure II-7, we see a three-stage shift register with the (nonlinear) feedback function $f(x_1, x_2, x_3) = x_2x_3$. From this, we can compute the next-state function, shown in the "successor table," for each of the eight states of the shift register. The eight states can accordingly be arranged in a state diagram, as shown. From the state diagram, we can directly observe the autonomous behavior of the device. For example, the initial state 4 (reading the states as three-digit binary numbers) leads to the sequence 4, 2, 1, 0, 0, 0, 0, 0, ...

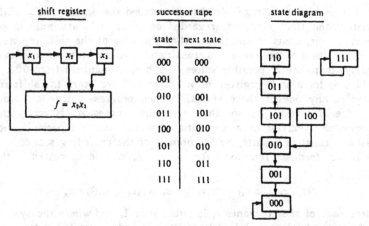

Fig. II-7. A three-stage shift register, and the succession of states.

A given shift register state has only two possible successors, since there is only one bit of uncertainty in the new term computed by the feedback function. Similarly, a given state has only two possible predecessors, since only one bit of the past has disappeared. The choice of any specific feedback function makes the assignment of successors unique, although predecessors need not be unique in general. For example, in Figure II-7, the state 2 has the predecessors 4 and 5, and the state 0 has the predecessors 0 and 1, whereas the states 4 and 6 have no predecessors at all. In Part Three of this book (Chapter VI, Section 1), the necessary and sufficient condition for unique predecessors is derived, and it is shown as a corollary that for *linear* shift registers, every state has exactly one predecessor. Graphically, this condition may be described as "pure cycles, without branches," from the appearance of the state diagram. In the terminology of classical physics, this is the "reversible" case, whereas cycles with branches correspond to *irreversible processes*. In the terminology of communication theory, the reversible case has also been called *information lossless*. Questions of reversibility or irreversibility are quite fundamental, and since they have been effectively answered for shift registers, it may be hoped that these answers can be extended to more general classes of finite state machines. We explore this possibility in the next section.

It is interesting to examine the superposition of the state diagrams for all possible shift registers of n stages. These diagrams, often referred to as de Bruijn diagrams because of N. G. de Bruijn's early paper [3], are shown in Figure II-8 for $n = 1, 2, 3, 4$, and in Figure II-9 for $n = 5$.

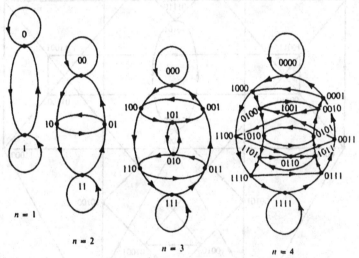

Fig. II-8. The de Bruijn diagrams for shift registers with 1, 2, 3, and 4 stages.

These diagrams are also the state diagrams for an nth order binary Markov process, that is, one in which the next term depends probabilistically on the preceding n terms. Thus, a *particular* n^{th} order Markov process would correspond to the de Bruijn diagram of order $n + 1$, with probabilities assigned to all the line segments in such a way that the probabilities assigned to the two segments directed out-

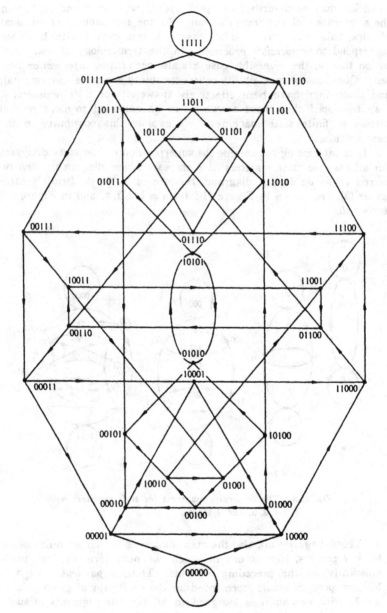

Fig. II-9. The de Bruijn diagram for $n = 5$.

ward from a given vertex sum to unity. Thus, a particular vertex (say 10110 in Figure II-9) would correspond to a previous history five terms into the past, and probabilities would be assigned to the two possible successors (01011 and 11011, in our example) corresponding to the probabilities that the next term produced by the process be a 0 or a 1.

Choosing a feedback function for our shift register actually specifies a rather trivial (because deterministic) Markov process, since the two segments emanating from a given vertex are assigned the probabilities 0 and 1 in some order. The truth table of the feedback function is thus a degenerate case of a probability table, in which only certainties have been allowed. (Probabilistic transitions between states, which seem to be the lifeblood of quantum mechanics, lie outside the domain of finite state machine theory.)

2.3. *Shift registers with input and output*

Figure II-10 shows a modification of the general shift register depicted in Figure II-6 to allow for the possibility of input and output. The effect of the shift register on the input sequence may be described as an *encoding*. Lest this term suggest exclusively cryptographic applications, it should be pointed out that devices of this sort are also central to the implementation of *efficient* codes. Specifically, the algebraic theory of error-correcting codes uses linear shift registers in this fashion for both encoding and decoding. (For details of this theory, see Peterson [29].)

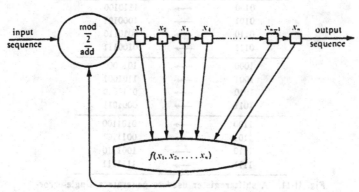

Fig. II-10. A general shift register with input and output.

As an example of the use of shift registers in the generation of error-correcting codes, we consider the famous (7, 4)-code, a single-error-correcting code of the Hamming variety, with sixteen ($=2^4$) binary codewords, each seven bits in length.   In Figure II-11, we see a diagram of the encoding device, and beneath it a table of input and output sequences.   The input sequences consist of four information bits which are "loaded" into the shift register, which is then allowed to start computing.   The corresponding output sequences are truncated after the first seven bits produced, thus yielding a seven-bit codeword for each four bits of information.   The value of this procedure is that the set of sixteen codewords thus generated has the property that any two of them differ in at least three of their seven places.   Hence, if an error occurs in one bit of a codeword, the resulting erroneous pattern is still closest to the *correct* codeword, differing from it in only one position, while differing from each of the other codewords in at least two places.   For this reason, the collection of codewords shown in Figure II-11 is known as a single-error-correcting code.

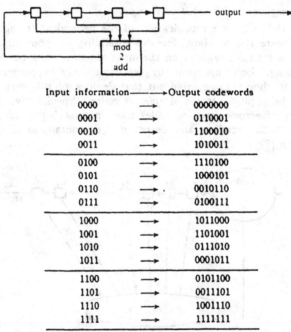

Input information		Output codewords
0000	→	0000000
0001	→	0110001
0010	→	1100010
0011	→	1010011
0100	→	1110100
0101	→	1000101
0110	→ .	0010110
0111	→	0100111
1000	→	1011000
1001	→	1101001
1010	→	0111010
1011	→	0001011
1100	→	0101100
1101	→	0011101
1110	→	1001110
1111	→	1111111

Fig. II-11.   A shift register used to generate a single-error-correcting code.

Fittingly enough, the decoding device, which accepts seven-bit words as input and produces four-bit words as output, on the assumption that the input is a codeword containing at most one error, is also a shift register device. If we regard the shift register in Figure II-11 as a linear operator, the decoder shift register is the inverse linear operator.

3. GENERALIZED SHIFT REGISTERS

3.1. *Binary n-stage machines*

Given any finite state machine with a specified number of states (say N), we can number the states from 0 to $N - 1$, using the binary numbering system and requiring only $n = \{\log_2 N\}$ binary digits for this representation, where the braces indicate that any fractional value is to be rounded *upward*. The transition from state to state can then be regarded as a "computation" into which the old state number and the present input number (since this too can be expressed by some finite number m of binary digits) enter as the independent variables, and the new state number is the dependent variable. More specifically, each binary digits of the new state number may be regarded as a Boolean function of $n + m$ binary variables, viz. the n digits $x_1, x_2, \ldots, x_n$ of the old state number and the m digits $y_1, y_2, \ldots, y_m$ of the input number. Furthermore, the output can be specified by some finite number r of binary digits, and we may regard each of these as a Boolean function of the same $n + m$ binary variables. Thus, the notion of a finite state machine, in its fullest generality, can be modeled by a set of $n + r$ Boolean functions, each of $n + m$ binary variables. A sketch of such a model is shown in Figure II-12.

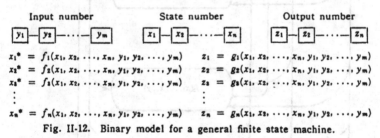

$$x_1{}^* = f_1(x_1, x_2, \ldots, x_n, y_1, y_2, \ldots, y_m) \qquad z_1 = g_1(x_1, x_2, \ldots, x_n, y_1, y_2, \ldots, y_m)$$
$$x_2{}^* = f_2(x_1, x_2, \ldots, x_n, y_1, y_2, \ldots, y_m) \qquad z_2 = g_2(x_1, x_2, \ldots, x_n, y_1, y_2, \ldots, y_m)$$
$$x_3{}^* = f_3(x_1, x_2, \ldots, x_n, y_1, y_2, \ldots, y_m) \qquad z_3 = g_3(x_1, x_2, \ldots, x_n, y_1, y_2, \ldots, y_m)$$
$$\vdots \qquad\qquad\qquad\qquad\qquad\qquad\qquad \vdots$$
$$x_n{}^* = f_n(x_1, x_2, \ldots, x_n, y_1, y_2, \ldots, y_m) \qquad z_n = g_n(x_1, x_2, \ldots, x_n, y_1, y_2, \ldots, y_m)$$

Fig. II-12. Binary model for a general finite state machine.

Although this is a valid description for *any* finite-state machine, the viewpoint is a natural generalization of the notion of a shift register. Indeed, instead of merely computing a new value for x_1, and shifting the other digits down the line, the situation in Figure II-12 indicates a wholesale recomputation of all the positions x_i. To specialize this situation to the one previously defined for shift registers proper, we have merely to impose the following conditions on the relationships in Figure II-12:

i. $m = r = 1$

ii. $x_1^* = f_1(x_1, x_2, \ldots, x_n, y) = f(x_1, x_2, \ldots, x_n) \oplus y$

iii. $x_{j-1}^* = x_j, \quad 1 \leq j \leq n - 1$

iv. $z = g(x_1, x_2, \ldots, x_n, y)$.

Of course, the state diagram for the general binary finite state machine is no longer constrained to the forms indicated in Figure II-8 and II-9. Much greater freedom of transition is possible in the general case. However, the *idea* of a state diagram, with permitted transitions, carries over completely.

Almost any reasonable special-purpose digital device can be put into the format shown in Figure II-12. Typical example include: (1) a standard binary counter, (2) a binary to BCD converter (or the inverse), (3) a Gray code counter (or converter), and even (4) a binary calculator, where the two operands may be written out as the input,

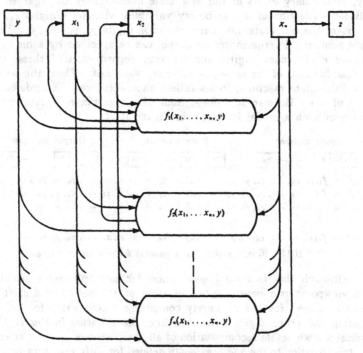

Fig. II-13.   Diagram of the general binary machine with restricted input and output.

the indicated operation is the state, and the answer is the output.

For the case of a single input bit, and a single output bit extracted from the state number itself, the model described in Figure II-12 appears as shown in Figure II-13.

3.2. *Autonomous behavior*

The situation depicted in Figure II-13 is even further simplified when we discuss only the autonomous behavior of a finite state machine, since at this point both input and output can be omitted entirely. The autonomous behavior of the device in Figure II-12 is fully specified, in principle, by the initial state and a set of n Boolean functions, each of n variables. Such a device is shown in Figure II-14.

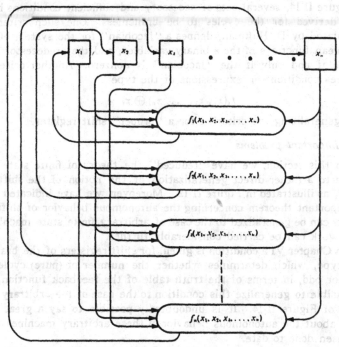

Fig. II-14. An autonomous n-state binary machine.

The effect of an autonomous binary machine is to assign, to each binary vector of length n, a *successor vector* of length n. In general, this is a many-to-one assignment, in that several different vectors may be given the same successor. In the important special case where the assignment of successors is a one-to-one mapping, the effect of the binary machine is to perform a *permutation* on the set of all 2^n binary vectors of length n.

Figure II-14 becomes an autonomous shift register when we impose the conditions $f_2 = x_1$; $f_3 = x_2$; ...; $f_n = x_{n-1}$. For this case, the necessary and sufficient condition that the machine perform a permutation on the binary vectors of length n is derived subsequently, in Chapter VI. The condition may be stated as follows: The expression $f_1(x_1, x_2, ..., x_n) \oplus x_n$ must in fact be independent of x_n.

As observed earlier, the case in which the assignment of successors to vectors is a permutation is also characterized by saying that *predecessors* of vectors are unique and by requiring that the state diagram consist of pure cycles, without branch points. For the device shown in Figure II-14, several sets of necessary and sufficient conditions have been derived for the cycles to be branchless. One such criterion, formulated by D. Huffman, defines a "Jacobian" for the system of the n Boolean functions of the n binary variables, and the predecessors are unique if and only if the "Jacobian" is nonzero. Another criterion involves conditions on expressions of the type

$$f_j(x_1, x_2, ..., x_n) \oplus x_i$$

thus generalizing the criterion for a "proper" shift register.

3.3. *Important problems*

In this section, we have "reduced" the theory of finite state machines to a rather direct generalization of the notion of the shift register, as illustrated in Figure II-12. Moreover, we have indicated that an important theorem concerning the autonomous behavior of shift registers can be generalized to the case of arbitrary finite state machines. This work can be carried considerably further.

In Chapter VI a condition is given, for shift registers of the branchless type, which determines whether the number of (pure) cycles is even or odd, in terms of the truth table of the feedback function. It is possible to generalize this condition to the case of the arbitrary machine of Figure II-14. It is undoubtedly possible to say a great deal more about the autonomous behavior of these arbitrary machines than has been done to date.

The other obvious area for additional research involves the interaction between input sequences (forcing functions) and the autonomous behavior of finite machines. The ultimate goal of such a theory is to predict the behavior of any finite state machine on the basis of its input sequence and its autonomous behavior.

PART TWO

THE LINEAR THEORY

Chapter III

SEQUENCES WITH RANDOMNESS PROPERTIES

1. INTRODUCTION

1.1. *Random sequences*

In a wide variety of situations arising in electronics, in digital computer work, in cryptography, and in numerous other fields, a need arises for random sequences. It is for such applications that tables of random numbers are compiled and ingenious noise-generating devices are designed.

In a great many of these applications, it would be very convenient to have a simple method of generating sequences which *look* random, even if upon closer inspection they can be shown to have certain regularities. In such instances, it is usually desired that the same apparatus, when used in the same way, will generate the same sequence. For example, if a random sequence generator is used in coding a message, the recipient of the message could then use an identical generator as a decoder. Specifically, if the sequence generator produces a sequence of 1's and − 1's, the message could be put in Morse Code, using 1's for dots and − 1's for dashes. The transmitter could then multiply each term of the Morse Code sequence by the corresponding term of the generator sequence. To decode, the receiver has only to multiply the received signal by the generated sequence once again to obtain the original Morse Code.

In the type of situation just described, it is desirable that the generator be as simple as possible, yet so constructed that slight modifications in the circuitry will enable it to generate new sequences, thus changing the code.

1.2. *Randomness characteristics*

In a sense, no finite sequence is ever truly random. The best that can be done is to single out certain properties as being associated with randomness, and to accept any sequence which has these properties as a *random sequence*.

24

In general, the sequences considered here will be binary sequences; that is, sequences of $+1$'s and -1's, corresponding to "on" and "off" in an electronic system.

The most familiar example of a random binary sequence arises from tossing an ideal coin consecutively, identifying heads as $+1$ and tails as -1. The following properties are then associated with randomness:

1. The number of heads is approximately equal to the number of tails.

2. Runs of consecutive heads or of consecutive tails frequently occur, with short runs being more frequent than long runs. More precisely, about one-half the runs have length 1, one-fourth have length 2, one-eighth have length 3, and so forth.

3. In addition to characteristics of the sort just noted, random sequences possess a special kind of auto-correlation function, peaked in the middle and tapering off rapidly at the ends.

1.3. *Auto-correlation*

If $\{a_n\} = \{a_0, a_1, a_2, \ldots\}$ is any sequence of real terms, its auto-correlation function $C(\tau)$ is defined as

$$C(\tau) = \lim_{N \to \infty} \frac{1}{N} \sum_{n=1}^{N} a_n a_{n+\tau}, \qquad (1)$$

provided that this limit exists. In particular, if $\{a_n\}$ is a *periodic* sequence with period p, this reduces to the *finite* sum

$$C(\tau) = \frac{1}{p} \sum_{n=1}^{p} a_n a_{n+\tau}. \qquad (2)$$

Here τ can be thought of as a phase shift of the sequence $\{a_n\}$. $C(\tau)$ measures the amount of similarity between the sequence and its phase shift. This is always highest for $\tau = 0$, and if $\{a_n\}$ is random, $C(\tau)$ is quite small for most other values of τ.

1.4. *The randomness postulates*

The ensuing chapters will discuss periodic binary sequences which have some or all of the following randomness characteristics.

R-1. In every period, the number of $+1$'s is nearly equal to the number of -1's. $\left(\text{More precisely, the disparity is not to exceed 1.}\right.$

Thus $\left| \sum_{n=1}^{p} a_n \right| \leq 1 .\Big)$

R-2. In every period, half the runs have length one, one-fourth have length two, one-eighth have length three, etc., as long as the number of runs so indicated exceeds 1.   Moreover, for each of these lengths, there are equally many runs of +1's and of −1's. [The *total* number of runs of + 1's equals the *total* number of runs of − 1's, because the runs of these two types alternate.]

  R-3.   The auto-correlation function $C(\tau)$ is two-valued. Explicitly

$$pC(\tau) = \sum_{n=1}^{p} a_n a_{n+\tau} = \begin{cases} p & \text{if } \tau = 0 \\ K & \text{if } 0 < \tau < p \end{cases} . \tag{3}$$

  This two-valuedness of the auto-correlation function is of considerable interest in itself.   It most frequently arises as the auto-correlation of a *pulse*, one of the least random of functions; thus it is all the more remarkable that *R*-3 can be satisfied by a sequence which also satisfies *R*-1 and *R*-2.

1.5. *Examples*

  Any sequence with the properties *R*-1, *R*-2 and *R*-3 will be called a *pseudo-noise* sequence.   (*Noise* is the standard designation in electronics for a random signal.)

  A typical example, with $p = 7$, is

  A: 1 1 1 −1 1 −1 −1 repeating periodically.   There are four +1's and three −1's, which is adequate to satisfy *R*-1.   Of the four runs, half have length one, and one-fourth have length two.   The auto-correlation is 1 in phase (true of any binary sequence), and −1/7 out of phase.

  The condition *R*-2 rather implies *R*-1; but otherwise, the three randomness postulates are independent.   To show this, the following examples can be given.

  B: 1 1 −1 1 1 1 −1 −1 −1 1 −1 has length $p = 11$, with six +1's and five −1's, so that *R*-1 is satisfied.   The auto-correlation satisfies *R*-3, the out-of-phase value being −1/11.   However, *R*-2 is violated by two runs of −1's of length one balanced by only the one run of +1's of length one.

  C: 1 1 1 1 1 −1 1 1 1 −1 −1 1 −1 is a sequence of length $p = 13$, which violates both *R*-1 and *R*-2.   Yet *R*-3 holds, with $K = + 1$.

  D: 1 −1 1 1 −1 −1 1 1 −1 satisfies *R*-1 and *R*-2, but fails completely on *R*-3.

  It will be shown in Section 5 that if the conditions *R*-1 and *R*-3 are met in the same sequence, then the out-of-phase value K is necessarily − 1.   Moreover, Section 5 will deal with the problem of determining *all* periodic sequences which satisfy *R*-3.

1.6. *Summary*

  There are many practical uses for random sequences.   For a binary

sequence, the following properties are associated with randomness:
  1. A balance of $+1$ and -1 terms.
  2. Two runs of length n for each run of length $n + 1$.
  3. A two-level auto-correlation function.
Any sequence with these properties (which are stated more precisely
in 1.4) is called a *pseudo-noise* sequence. The generation of pseudo-
noise sequences, and their additional characteristics, provides the basic
subject matter of the remaining Sections.

2. SHIFT REGISTER SEQUENCES

2.1. *Shift registers with feedback*

  A *shift register* is an arrangement of r tubes in a row (Fig. III-1),
each tube either "on" (1) or "off" (0), which shifts the contents of

$$\square \rightarrow \square \rightarrow \square \rightarrow \square \rightarrow \square$$

Fig. III-1.

each tube to the next tube, in time with a clock pulse. If no new
signal is introduced into the first tube during this process, then at
the end of r shifts (or even sooner), all the tubes will be "off", and
will remain that way.

  One way to keep the shift register active is to feed back the states
of certain of the r tubes into the first tube. That is, leads from certain
of the r tubes feed into a mod 2 adder, and when the next shift of
the register occurs, the mod 2 sum is transferred to the first tube (see
Fig. III-2).

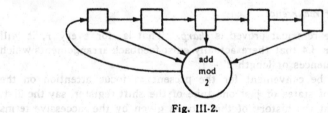

Fig. III-2.

  Addition modulo 2 (or "mod 2") is defined by the Table

+ (mod 2)	0	1
0	0	1
1	1	0

(1)

In general, odd integers are replaced by 1, and even integers by 0. (If one is accustomed to thinking of on as -1, and off as $+1$, then the same effect is achieved by using a multiplier instead of a mod 2 adder.)

Henceforth, the term *shift register* will always refer to a shift register with feedback, as in Fig. III-2.

2.2. *Periodicity of shift register states*

It is not hard to show that the succession of states in a shift register is periodic.
In fact:

> *Theorem 2.1.*   The succession of states in a shift register
> is periodic, with a period $p \leqq 2^r - 1$, where r is the number
> of tubes.

Proof. Each state of the shift register is completely determined by the previous state. Hence, if it *ever* happens that a state is the same as some earlier state, then the following states are the same, so that a periodicity is established.

With r tubes in a shift register, each one either "on" or "off", there are only 2^r possible states. Thus a repetition occurs somewhere among the first $2^r + 1$ states, and there is periodicity, with $p \leqq 2^r$.

Finally, if the state "all 0's" ever occurs, the subsequent states will also consist of "all 0's", and the periodicity is $p = 1$. Thus a *long* period cannot include this state; and $p \leqq 2^r - 1$.

Note that Theorem 2.1 holds no matter what the *initial* state of the shift register is.

2.3. *Linear recurrence*

The theorem just proved is *sharp*. That is, for every r, it will be shown in 3.4 that there actually exist feedback arrangements which lead to sequences of length $2^r - 1$.

It will be convenient for the present to focus attention on the succession of states of just one tube of the shift register, say the first. Suppose that the history of this tube is given by the successive terms $a_0, a_1, a_2, \ldots, a_n$. From the feedback arrangement, a_n is a sum (mod 2) of the contents of several of the tubes at the $(n-1)^{st}$ state. In fact, every tube in the mod 2 sum for a_n can ultimately be traced back to a previous state of the first tube itself. That is,

> *Theorem 2.2.* a_n satisfies an equation of the type

$$a_n = c_1 a_{n-1} + c_2 a_{n-2} + \cdots + c_r a_{n-r} \qquad (2)$$

where the coefficients $c_1, c_2, \ldots, c_r$ are all 1's and 0's and do not depend on n. (Addition, of course, is modulo 2.) Such a relationship is called a *linear recurrence*, and any sequence $\{a_n\}$ which satisfies (2) is called a *linear recurring sequence*.

2.4. Example

As an example of Theorems 2.1 and 2.2, consider the three-tube shift register of Fig. III-3. If the initial state is 1 1 1 then the succession of states is:

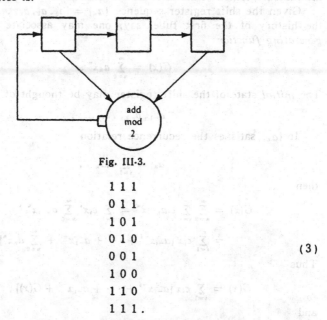

Fig. III-3.

$$
\begin{array}{ccc}
1 & 1 & 1 \\
0 & 1 & 1 \\
1 & 0 & 1 \\
0 & 1 & 0 \\
0 & 0 & 1 \\
1 & 0 & 0 \\
1 & 1 & 0 \\
1 & 1 & 1 .
\end{array}
\tag{3}
$$

The eighth state is the same as the first, so that the period is $7 = 2^3 - 1$, the longest period allowed by Theorem 2.1. Indeed, every possible state except 0 0 0 has occurred.

The sequence of states of the *first* tube, 1010011, satisfies the linear recurrence

$$ a_n = a_{n-1} + a_{n-3} . \tag{4} $$

Next, it should be observed that the following two facts hold in general, as well as for the example just given.

1. The history of the second tube is the history of the first tube, except for a delay of one state. Similarly for the other tubes.

2.   Hence, each tube satisfies the same linear recurrence relation as the first.   Thus the *entire shift register* can be thought of as satisfying the recurrence relation.

2.5.   *Generating functions*

There are two basically different methods for studying recurring sequences.   If a single tube is considered, the method of generating functions, introduced in this section, yields quick results.   If the entire shift register is thought of as satisfying the recurrence relation, then the Matrix Method of 2.9 is to be recommended.

Given the shift register sequence $\{a_n\} = \{a_0, a_1, a_2, \ldots\}$ describing the history of the first tube, say, one may associate with it the *generating function*

$$G(x) = \sum_{n=0}^{\infty} a_n x^n. \tag{5}$$

The *initial* state of the shift register may be thought of as

$$a_{-1}, a_{-2}, \ldots, a_{-r}. \tag{6}$$

If $\{a_n\}$ satisfies the recurrence relation

$$a_n = \sum_{i=1}^{r} c_i a_{n-i},$$

then

$$G(x) = \sum_{n=0}^{\infty} \sum_{i=1}^{r} c_i a_{n-i} x^n = \sum_{i=1}^{r} c_i x^i \sum_{n=0}^{\infty} a_{n-i} x^{n-i}$$

$$= \sum_{i=1}^{r} c_i x^i [a_{-i} x^{-i} + \cdots + a_{-1} x^{-1} + \sum_{n=0}^{\infty} a_n x^n].$$

Thus

$$G(x) = \sum_{i=1}^{r} c_i x^i [a_{-i} x^{-i} + \cdots + a_{-1} x^{-1} + G(x)],$$

and

$$G(x) - \sum_{i=1}^{r} c_i x^i G(x) = \sum_{i=1}^{r} c_i x^i (a_{-i} x^{-i} + \cdots + a_{-1} x^{-1}).$$

In other words,

$$G(x) = \frac{\sum_{i=1}^{r} c_i x^i (a_{-i} x^{-i} + \cdots + a_{-1} x^{-1})}{1 - \sum_{i=1}^{r} c_i x^i}. \tag{7}$$

This expresses $G(x)$ entirely in terms of the *initial conditions* $a_{-1}, a_{-2}, \ldots, a_{-r}$, and the *feedback coefficients* $c_1, c_2, \ldots, c_r$.   In fact, the *denominator* in (7) is independent even of the initial conditions.

With the initial conditions $a_{-1} = a_{-2} = \cdots = a_{1-r} = 0$, $a_{-r} = 1$, expression (7) reduces to

$$G(x) = \frac{c_r}{1 - \sum_{i=1}^{r} c_i x^i} . \qquad (8)$$

It will be convenient to refer to the rth degree polynomial

$$f(x) = 1 - \sum_{i=1}^{r} c_i x^i , \qquad (9)$$

as the *characteristic polynomial* of the sequence $\{a_n\}$ and of the shift register which produced it.

Note that the characteristic polynomial of a shift register is obtained by taking $f(x) = 1 - \sum x^j$, where the sum is taken over those values of j for which the jth tube feeds back into the mod 2 adder (Fig. III-2).

*2.6. *Recurring sequences mod 0*

So far, the recurrence relation has been used to determine $G(x)$, but no mention of the fact that addition is modulo 2 has been made. Thus, the formulas (7), (8) for $G(x)$ hold for ordinary addition as well as for addition modulo 2.

The *Fibonacci sequence* is defined by the initial conditions $a_0 = 0$, $a_{-1} = 1$, and the recursion relation $a_n = a_{n-1} + a_{n-2}$, for $n > 0$. Thus the sequence begins

$$1\ 1\ 2\ 3\ 5\ 8\ 13\ 21\ 34\ \cdots \qquad (10)$$

According to (8), the Fibonacci sequence should be the coefficients in the power series

$$G(x) = \frac{1}{1 - x - x^2} \qquad (11)$$

Ordinary long division shows indeed that

$$G(x) = 1 + x + 2x^2 + 3x^3 + 5x^4 + 8x^5 + \cdots \qquad (12)$$

The question of convergence can be brought into play. A power series converges in the largest circle about the origin which is free from singularities. Now, $G(x) = 1/(1 - x - x^2)$ has singularities wherever the characteristic polynomial $f(x) = 1 - x - x^2$ has roots. The roots of $1 - x - x^2 = 0$ are $x = (-1 \pm \sqrt{5})/2$. Of these, $x = (\sqrt{5} - 1)/2$ is the closer to the origin. This, then, is the radius of convergence R of the power series $G(x)$. On the other hand, a well-known formula for the *radius of convergence* says that

$$\frac{1}{R} = \overline{\lim_{n \to \infty}} \frac{a_{n+1}}{a_n} . \tag{13}$$

For the Fibonacci sequence, this becomes

$$\lim_{n \to \infty} \frac{a_{n-1}}{a_n} = \frac{2}{\sqrt{5} - 1} = \frac{2(\sqrt{5} + 1)}{5 - 1} = \frac{1 + \sqrt{5}}{2} . \tag{14}$$

That is, the limiting ratio of consecutive terms in Fibonacci's sequence is $(1 + \sqrt{5})/2 = 1.618. \ldots$

Working mod 2, the radius of convergence of $G(x)$ is no longer defined. However, the *roots* of the characteristic polynomial, and the closely related question of the *factorization* of the characteristic polynomial, are still the principal methods of obtaining information about the sequence itself.

2.7. *Determining the period*

Theorem 2.1 shows that shift register sequences are periodic, and gives the *upper bound* for the period. Equation (8) makes it possible to give a more exact description.

> *Theorem* 2.3: If an r-tube shift register sequence $A = \{a_n\}$ obeys the initial conditions $a_{-1} = a_{-2} = \cdots = a_{1-r} = 0$, $a_{-r} = 1$, then the period of A is the smallest positive integer p for which the characteristic polynomial $f(x)$ divides $1 - x^p$, modulo 2.

Proof: Under the initial conditions stated,

$$G(x) = \frac{1}{f(x)} = \sum_{n=0}^{\infty} a_n x^n . \tag{15}$$

i. If A has period p, then

$$\frac{1}{f(x)} = (a_0 + a_1 x + \cdots + a_{p-1} x^{p-1}) + x^p(a_0 + a_1 x + \cdots + a_{p-1} x^{p-1})$$
$$+ x^{2p}(a_0 + a_1 x + \cdots + a_{p-1} x^{p-1}) + \cdots$$
$$= (a_0 + a_1 x + \cdots + a_{p-1} x^{p-1})(1 + x^p + x^{2p} + x^{3p} + \cdots)$$
$$= (a_0 + a_1 x + \cdots + a_{p-1} x^{p-1})/(1 - x^p) .$$

Thus

$$f(x)[a_0 + a_1 x + \cdots + a_{p-1} x^{p-1}] = 1 - x^p , \tag{16}$$

and $f(x)$ divides $1 - x^p$.

ii. Conversely, if $f(x)$ divides $1 - x^p$, let the quotient be

$$\alpha_0 + \alpha_1 x + \cdots + \alpha_{p-1} x^{p-1} .$$

Then

$$\frac{1}{f(x)} = \frac{\alpha_0 + \alpha_1 x + \cdots + \alpha_{p-1}x^{p-1}}{1 - x^p}$$

$$= (\alpha_0 + \alpha_1 x + \cdots + \alpha_{p-1}x^{p-1})(1 + x^p + x^{2p} + \cdots)$$

$$= (\alpha_0 + \alpha_1 x + \cdots + \alpha_{p-1}x^{p-1}) + x^p(\alpha_0 + \alpha_1 x + \cdots + \alpha_{p-1}x^{p-1})$$

$$+ \cdots = G(x) = \sum_{n=0}^{\infty} a_n x^n,$$

and equating coefficients of like powers of x, $\{a_n\} = \{\alpha_n\}$, so that A has p or some factor thereof as its period.

Thus, the *smallest* positive integer p for which $f(x)$ divides $1 - x^p$ is the period of A.

> *Corollary.* By (7), $G(x) = g(x)/f(x)$, where the numerator $g(x)$ has degree less than the degree of $f(x)$. (If the shift register is really making use of the fact that it has r tubes, then $f(x)$ has degree r. It is no loss of generality to assume that this is always the case.) If $g(x)$ has no factors in common with $f(x)$, the proof of Theorem 2.3 still works, and the smallest p such that $f(x)$ divides $1 - x^p$, called the *exponent* of $f(x)$, is the period of the corresponding sequence. When $g(x) = 1$, this is simply Theorem 2.3 itself. Another very important case is when $f(x)$ is *irreducible*, in which case it can have no factors in common with $g(x)$, a polynomial of lower degree, unless $g(x) = 0$, which corresponds to the initial condition "all 0's". Thus, when $f(x)$ is irreducible, the period of the shift register sequence does not depend on the initial conditions, excepting only the initial condition "all 0's".

2.8. *A necessary condition for maximum length*

Definition. A sequence generated by an r-tube shift register will be said to have *maximum length* if its period is $p = 2^r - 1$.

> *Theorem 2.4.* If the sequence A has maximum length, its characteristic polynomial is irreducible (that is, cannot be factored).

Proof. Since A runs through $2^r - 1$ terms before it repeats, every sequence of 1's and 0's of length r (except r consecutive 0's) can be found in A. In particular, somewhere there is a 1 followed by $r - 1$ 0's. Beginning there, the initial condition of Theorem 2.3 is satisfied. Consequently, the period of A is the *exponent* of $f(x)$.

Suppose $f(x)$ factors: $f(x) = s(x) \cdot t(x)$. Then $1/f(x) = \alpha(x)/s(x) +$

$\beta(x)/t(x)$ by partial fraction decomposition.   Suppose $s(x)$ and $t(x)$ have degrees $r_1 > 0$ and $r_2 > 0$, respectively, with $r_1 + r_2 = r$. $\alpha(x)/s(x)$ is then a power series whose coefficients are periodic with period at most $2^{r_1} - 1$, and $\beta(x)/t(x)$ is a power series whose coefficients have period at most $2^{r_2} - 1$.   Then the sum $1/f(x) = \alpha(x)/s(x) + \beta(x)/t(x)$ represents a power series whose coefficients have period at most the *least common multiple* of the individual periods, which in turn cannot exceed the *product* of the periods.   Thus

$$2^r - 1 \leqq (2^{r_1} - 1)(2^{r_2} - 1) = 2^{r_1 + r_2} - 2^{r_1} - 2^{r_2} + 1$$
$$\leqq 2^r - 2 - 2 + 1 = 2^r - 3. \tag{17}$$

This contradiction shows that the assumption "$f(x)$ factors" is invariably false.

In the proof of this Theorem, the partial fraction decomposition of $f(x)$ assumes that $s(x)$ and $t(x)$ are *distinct* factors.   For a mathematically complete demonstration, it should be observed that if $f(x) = s^2(x)$, then the period of $f(x)$ is twice that of $s(x)$, hence, at most $2(2^{r/2} - 1)$ $< 2^r - 1$.   In this manner, the case of *repeated* factors can also be dealt with.

The converse of Theorem 2.4 does not hold.   There actually exist irreducible polynomials which correspond to no maximum-length sequences.   For example, $f(x) = x^4 + x^3 + x^2 + x + 1$ is irreducible, but since it divides $1 - x^5$, Theorem 2.3 asserts that sequences corresponding to it have period 5, rather than the "maximum" period $2^4 - 1 = 15$.

Similarly, $f(x) = x^6 + x^3 + 1$ is irreducible, but has exponent 9 rather than 63.

Distinguishing between irreducible and reducible polynomials and selecting the "maximum exponent" polynomials from among the irreducible ones, is the principal task of Section 3.   It is not until Section 4 that the maximum-length shift register sequences will be shown to satisfy the three "randomness postulates" of Section 1.

*2.9.   The matrix method

Each state of an r-tube shift register can be thought of as an r-dimensional vector.   The shift register is then a linear operator which changes each state into the next.   It is a familiar fact that a linear operator, operating on r-dimensional vectors, is most conveniently represented by an $r \times r$ matrix.

As an example, the 3-tube shift register of Section 2.4 is mathematically *equivalent* to the matrix

$$M = \begin{pmatrix} 1 & 1 & 0 \\ 0 & 0 & 1 \\ 1 & 0 & 0 \end{pmatrix}.$$

In particular,

$$(1\ 1\ 1)\begin{pmatrix}1&1&0\\0&0&1\\1&0&0\end{pmatrix} = (0\ 1\ 1),$$

$$(0\ 1\ 1)\begin{pmatrix}1&1&0\\0&0&1\\1&0&0\end{pmatrix} = (1\ 0\ 1),$$

and so forth.

In general, a shift register matrix takes the form

$$M = \begin{pmatrix}c_1&1&0&\cdots&0\\c_2&0&1&&0\\ \vdots&\vdots&&\ddots&\vdots\\c_{r-1}&0&0&&1\\c_r&0&0&\cdots&0\end{pmatrix}, \qquad (18)$$

with 1's along the diagonal *above* the main diagonal, and the "feedback coefficients" down the first column. No proof is needed here except explicit verification:

$$(a_{n-1}, a_{n-2},\ldots,a_{n-r})\begin{pmatrix}c_1&1&0&\cdots&0\\c_2&0&1&\cdots&0\\ \vdots&\vdots&&\ddots&\vdots\\ & & & &1\\c_r&0&0&\cdots&0\end{pmatrix}$$

$$= (c_1 a_{n-1} + c_2 a_{n-2} + \cdots + c_r a_{n-r}, a_{n-1}, a_{n-2},\ldots,a_{n-r+1})$$

$$= (a_n, a_{n-1},\ldots,a_{n-r+1}).$$

The *characteristic equation* of the matrix M is

$$f(\lambda) = det|M - \lambda I| = \begin{vmatrix}c_1-\lambda&1&0&\cdots&0\\c_2&-\lambda&1&\cdots&0\\c_3&0&-\lambda&\cdots&0\\ \vdots&\vdots&\vdots&\ddots&\vdots\\c_r&0&0&&-\lambda\end{vmatrix}$$

$$= (-\lambda)^{r-1}(c_1 - \lambda) - c_2(-\lambda)^{r-2} + c_3(-\lambda)^{r-3}\cdots(-1)^r c_r$$

$$= -(-\lambda)^r\left[1 - \frac{c_1}{\lambda} - \frac{c_2}{\lambda^2} - \frac{c_3}{\lambda^3} - \cdots - \frac{c_r}{\lambda^r}\right]$$

$$= \frac{(-1)^{r+1}}{x^r}[1 - (c_1 x + c_2 x^2 + \cdots + c_r x^r)],$$

where the substitution $x = 1/\lambda$ has been made. Thus, except for the factor of $(-1)^{r+1} x^r$, the characteristic equation of M is the characteristic polynomial of the shift register.

*2.10. *Periodicity and the characteristic equation*

Determining the periodicity of the shift register is equivalent, except in degenerate cases, to finding the lowest power p of the equivalent matrix M such that $M^p = I$, the identity matrix.

A well-known theorem of matrix theory asserts that every matrix formally satisfies its characteristic equation: thus $f(M) = 0$. If $f(x)$ divides $1 - x^p$, then M is a root of $I - X^p = 0$, since it is a root of the factor $f(X) = 0$. That is, if $f(x)$ divides $1 - x^p$, then $M^p = I$. Conversely, if $f(x)$ is irreducible, it divides *every* polynomial which has the root M in common with it, and will divide $1 - x^p$ if $M^p = I$.

This is a matrix theory proof of Theorem 2.3, at least in the important case when $f(x)$ is irreducible. This part of the theory can be done entirely by matrix theory, and is frequently done so in the literature.

*2.11. *The extension to mod m*

Anywhere in this chapter where "mod 2" has been used, a similar discussion applies to the more general "mod m" case. Physically, this corresponds to a shift register with r tubes, each capable of m states (e.g., intensities).

Such a shift register will have period $m^r - 1$ or less. If "maximum length" is defined as $p = m^r - 1$, and "modulo 2" is replaced by "modulo m" wherever it appears, then Theorems 2.3 and 2.4 remain true in their entirety.

2.12. *Summary*

One rather simple method of generating binary sequences is to use a shift register with a mod 2 feedback. The sequences so generated are periodic, with periods not exceeding $2^r - 1$, where r is the number of tubes of the shift register.

Every shift register sequence $\{a_n\}$ satisfies a linear recurrence relation $a_n = \sum_{i=1}^{r} c_i a_{n-i}$, where c_i is 1 or 0 according to whether the ith tube is or is not involved in the feedback circuit. By the method of generating functions, the following facts can be shown:

1. $$\sum_{n=0}^{\infty} a_n x^n = \frac{g(x)}{f(x)},$$

where the numerator is a polynomial of degree less than r, and the denominator is the *characteristic polynomial* $f(x) = 1 - \sum_{i=1}^{r} c_i x^i$.

2. The *period* of the shift register sequence is the smallest positive integer p for which $1 - x^p$ is divisible by $f(x)$.

3. A necessary (but insufficient) condition for the period of $\{a_n\}$ to be $2^r - 1$ is that $f(x)$ be irreducible (i.e., unfactorable).

The results thus obtained by generating functions can also be derived by matrix theory, for every shift register is mathematically equivalent to an $r \times r$ matrix. The characteristic equation of this matrix is essentially the same as the characteristic polynomial of the shift register.

3. POLYNOMIALS MODULO 2

3.1. *Mersenne prime periods*

Every irreducible polynomial mod 2, of degree $r > 1$, divides the polynomial $1 - x^{2^r-1}$ [39]. When $r = 3$, for example, this fact asserts that both $1 + x + x^3$ and $1 + x^2 + x^3$ divide $1 - x^7$. Actually, $1 - x^7 = (1 - x)(1 + x + x^3)(1 + x^2 + x^3)$ modulo 2.

From this fact and from Theorem 2.3, it is simple to conclude:

> *Theorem 3.1.* If a sequence has an irreducible characteristic polynomial of degree r, the period of the sequence is a *factor* of $2^r - 1$.

> *Corollary.* If $2^r - 1$ is prime, every irreducible polynomial of degree r corresponds to a shift register sequence of maximum length.

Indeed, in this case, the only factor of $2^r - 1$ is $2^r - 1$ itself. When $2^r - 1$ is prime, it is known in the literature as a *Mersenne prime*. The first twenty-three Mersenne primes have been computed, and are listed for reference in Table III-1.

Table III-1. Mersenne primes: $2^p - 1$, with $p =$

2	17	107	2203	9689
3	19	127	2281	9941
5	31	521	3217	11213
7	61	607	4253	
13	89	1279	4423	

3.2. *Two number-theoretic functions*

There is an explicit formula for the number of irreducible polynomials mod 2 of degree r, and even for the number of "maximum

exponent'' polynomials of degree r. These formulas involve two functions used commonly in Number Theory, which are defined herewith.

By the Unique Factorization Theorem of arithmetic, every integer
$n > 1$ is a product of powers of distinct primes,

$$n = \prod_{i=1}^{k} p_i^{a_i}. \tag{1}$$

In terms of this factorization, the Euler φ-function is defined by

$$\phi(n) = \begin{cases} 1 & \text{if } n = 1 \\ \prod_{i=1}^{k} p_i^{a_i-1}(p_i - 1) & \text{if } n > 1. \end{cases} \tag{2}$$

In particular, if P denotes a prime, $\phi(P) = P - 1$. If Q is also a

Table III-2.

n	$\mu(n)$	$\phi(n)$	n	$\mu(n)$	$\phi(n)$
1	1	1	26	1	12
2	−1	1	27	0	18
3	−1	2	28	0	12
4	0	2	29	−1	28
5	−1	4	30	−1	8
6	1	2	31	−1	30
7	−1	6	32	0	16
8	0	4	33	1	20
9	0	6	34	1	16
10	1	4	35	1	24
11	−1	10	36	0	12
12	0	4	37	−1	36
13	−1	12	38	1	18
14	1	6	39	1	24
15	1	8	40	0	16
16	0	8	41	−1	40
17	−1	16	42	−1	12
18	0	6	43	−1	42
19	−1	18	44	0	20
20	0	8	45	0	24
21	1	12	46	1	22
22	1	10	47	−1	46
23	−1	22	48	0	16
24	0	8	49	0	42
25	0	20	50	0	20

prime, $\phi(PQ) = (P - 1)(Q - 1)$. Also, $\phi(P^2) = P(P - 1)$. The ϕ-function is tabulated in Table III-2.

An equivalent definition is: $\phi(n)$ is the number of positive irreducible fractions not greater than 1, with denominator n.

For example, if $n = 6$, the only irreducible sixths are 1/6 and 5/6. Thus $\phi(6) = 2$. This agrees with the previous definition, by which $\phi(6) = \phi(2\cdot3) = (2 - 1)(3 - 1) = 2$.

The Möbius μ-function is defined from (1) by

$$\mu(n) = \begin{cases} 1 & \text{if } n = 1 \\ 0 & \text{if } \prod_{i=1}^{k} a_i > 1 \\ (-1)^k & \text{otherwise (i.e., if } n \text{ is the product of } \\ & k \text{ distinct primes)}. \end{cases} \tag{3}$$

If P is a prime, $\mu(P) = -1$. If Q is also a prime, $\mu(PQ) = +1$, while $\mu(P^2) = 0$. The Möbius function is tabulated for $n \le 50$ in Table III-2.

*3.3. *Some interesting relationships*

The Euler and Möbius functions have many other interesting properties. For example,

$$\sum_{d|n} \phi(d) = n, \tag{4}$$

where the sum is extended over all positive divisors d of n.

Analogously,

$$\sum_{d|n} \mu(d) = \begin{cases} 1 & \text{if } n = 1 \\ 0 & \text{if } n > 1. \end{cases} \tag{5}$$

Also,

$$\frac{\phi(n)}{n} = \sum_{d|n} \frac{\mu(d)}{d}. \tag{6}$$

Finally, there are always $\phi(n)$ primitive n^{th} roots of unity (cf. Sec. 5), and the sum of these roots is exactly $\mu(n)$.

3.4. *The number of irreducible polynomials*

The number of irreducible polynomials modulo 2 of degree r is given by

$$\Psi_2(r) = \frac{1}{r} \sum_{d|r} 2^d \mu\left(\frac{r}{d}\right). \tag{7}$$

where the sum is extended over all positive divisors d of r.

For example, to compute the number of irreducible polynomials mod 2 of degree 8, the divisors of 8 are $d = 1, 2, 4, 8$. Then $r/d = 8$, 4, 2, 1 and $\mu(8'd) = 0, 0, -1, +1$. Thus, $\sum\limits_{d|8} 2^d \mu(8/d) = 0 + 0 - 16 + 256 = 240$, and

$$\Psi_2(8) = \frac{1}{8} \sum_{d|8} 2^d \mu\left(\frac{8}{d}\right) = 30 .$$

The values of $\Psi_2(r)$ are included in Table III-3 for $r \leqq 24$.

It has already been observed that not all the irreducible polynomials of degree r have maximum exponent. That is, they do not all correspond to shift register sequences of length $2^r - 1$. The number of polynomials mod 2 of degree r which have maximum exponent is given by

$$\lambda_2(r) = \frac{\phi(2^r - 1)}{r} . \tag{8}$$

Table III-3.

r	$2^r - 1$	$\lambda_2(r)$	$\Psi_2(r)$	$3^r - 1$	$\lambda_3(r)$	$\Psi_3(r)$
1	1	1	2	2	1	3
2	3	1	1	8	2	3
3	7	2	2	26	4	8
4	15	2	3	80	8	18
5	31	6	6	242	22	48
6	63	6	9	728	48	116
7	127	18	18	2,186	156	312
8	255	16	30	6,560	320	810
9	511	48	56	19,682	1,008	2,457
10	1,023	60	99	59,048	2,640	5,880
11	2,047	176	186	177,146	7,700	16,104
12	4,095	144	335	531,440	13,824	44,220
13	8,191	630	630	1,594,322	61,320	122,640
14	16,383	756	1,161	4,782,968	170,352	341,484
15	32,767	1,800	2,182			
16	65,535	2,048	4,080			
17	131,071	7,710	7,710			
18	262,143	8,064	14,532			
19	524,287	27,594	27,594			
20	1,048,575	24,000	52,377			
21	2,097,151	84,672	99,858			
22	4,194,303	120,032	190,557			
23	8,388,607	356,960	364,722			
24	16,777,215	276,480	698,870			

For example, the number of maximum exponent polynomials of degree 8 is

$$\lambda_2(8) = \frac{\phi(255)}{8} = \frac{\phi(3.5.17)}{8} = \frac{2.4.16}{8} = 16 \ .$$

Thus, in this case, barely half of the irreducible polynomials have a maximum exponent. However, when $2^r - 1$ is prime, $\Psi_2(r) = \lambda_2(r) = (2^r - 2)/r$.

A tabulation of $\lambda_2(r)$ also appears in Table III-3.

3.5. *Generalization to other moduli*

The formulas of the previous section generalize to *any* prime modulus p. Thus, the number of irreducible polynomials of degree r mod p is

$$\Psi_p(r) = \frac{1}{r} \sum_{d \mid r} p^d \mu\left(\frac{r}{d}\right), \tag{9}$$

and the number which have *maximum exponent* (in this case $p^r - 1$ is the maximum index) is

$$\lambda_p(r) = \frac{\phi(p^r - 1)}{r} \ . \tag{10}$$

It can "never" happen, for $p > 2$, that $p^r - 1$ is prime, since this number is always even. (Trivial exception: $3^1 - 1$ is prime.) Thus the interesting case of Mersenne primes has no true analog for prime moduli other than 2.

*3.6. *Shorter periods*

It is not difficult to catalog the possible shorter periods (i.e., smaller exponents) which irreducible polynomials mod 2 of degree r may possess. It is also easy to state *how many* irreducible polynomials there are with each of these exponents.

> *Theorem 3.2.* Every factor f of $2^r - 1$ which is *not* a factor of any number $2^s - 1$ with $s < r$ occurs as the exponent of irreducible polynomials of degree r. Precisely, there are $\phi(f)/r$ irreducible polynomials of degree r with exponent f.

Example. When $r = 8$, $2^r - 1 = 255$, which has the factors 1, 3, 5, 15, 17, 51, 85, 255. Of these, 1, 3, 5, and 15 already divide $2^4 - 1$. But the others all occur as "periods" for irreducible polynomials.

In particular, there are $\phi(17)/8 = 2$ with exponent 17, $\phi(51)/8 = 4$ with exponent 51, $\phi(85)/8 = 8$ with exponent 85, and $\phi(255)/8 = 16$ with exponent 255. Thus, the 30 irreducible polynomials of degree 8 are all accounted for $(2 + 4 + 8 + 16 = 30)$.

*3.7. *The reducible case*

The theorems and formulas of this section have been presented without proof. Most of the proofs, as well as the interrelation between (7) and (8), are given in Section 5.

For the sake of completeness, two theorems will now be stated which determine the periods of shift register sequences when the characteristic polynomial is *reducible*.

> *Theorem 3.3.* If $f(x) = s(x)t(x)$, where $s(x)$ has no factor in common with $t(x)$, then the exponent of $f(x)$ is the least common multiple of the exponents of $s(x)$ and $t(x)$.

This can be proved either directly, from the definition of "exponent"; or else in analogous fashion to the proof of Theorem 2.4.

> *Theorem 3.4.* If $f(x)$ is irreducible, and $g(x) = [f(x)]^n$, the period q of $g(x)$ is a multiple $e(n)$ of the period p of $f(x)$. That is, $q = e(n)p$, where $e(n)$ is given by

n	$e(n)$	n	$e(n)$	
1	1	5	8	
2	2	6	8	(11)
3	4	7	8	
4	4	8	8	etc.

Thus, to find the period (i.e., the exponent) of *any* polynomial corresponding to a shift register, it can first be factored into powers of *irreducible* polynomials. The period of each factor is found using Theorem 3.4, and then the period of the product is found by using Theorem 3.3.

3.8. *Summary*

In the special case that $2^r - 1$ is a prime, every irreducible polynomial of degree r has exponent $2^r - 1$; but this fails if $2^r - 1$ is composite.

In terms of two functions of elementary number theory, it is possible to give explicit formulas for the number of irreducible polynomials of degree r, and the number of maximum exponent polynomials of degree r. From these formulas, it can be seen that

1. Polynomials of both types exist for every positive r.

2. The number of polynomials is the same for both types if and only if $2^r - 1$ is prime.

3. The number of irreducible polynomials is slightly less (but asymptotically equal) to $2^r/r$. The number of maximum exponent polynomials is on the same order of magnitude.

The results obtained are generalized to other moduli; extended to cover irreducible polynomials with non-maximum exponent; and applied to the case of *reducible* polynomials.

4. RANDOMNESS PROPERTIES OF SHIFT REGISTER SEQUENCES

4.1. *Randomness postulate R − 1*

Suppose an r-tube shift register goes through all $2^r - 1$ possible states before it repeats. Reading "on" as 1 and "off" as 0, these states are the integers from 1 to $2^r - 1$, in binary notation. The corresponding maximum-length shift register sequence $\{a_n\}$ can be thought of as the *units digit* (in the binary scale) of the numbers from 1 to $2^r - 1$. The units digit is 1 for odd numbers and 0 for even numbers. From 1 to $2^r - 1$, there are 2^{r-1} odd numbers, and $2^{r-1} - 1$ even numbers. Thus, in any maximum-length shift register sequence, there are 2^{r-1} ones and $2^{r-1} - 1$ zeros. That is:

Theorem 4.1. The randomness property $R - 1$ holds for all maximum-length shift register sequences.

Strictly speaking, $R - 1$ only applies to sequences of $+ 1$'s and $- 1$'s. However, it is perfectly straightforward to interpret "off" as $+ 1$ and "on" as $- 1$, in which case $R - 1$ holds as originally stated. A similar remark applies to 4.2, and the postulate $R - 2$.

4.2. *Randomness postulate R − 2*

In the maximum-length shift register sequence $\{a_n\}$, there are $2^r - 1$ ways to choose r consecutive terms. That is, every possible array of r consecutive terms, except all 0's, occurs exactly once.

In particular, r consecutive 1's occurs once. This run of 1's must be preceded by 0 and followed by 0, or there would be other runs of r consecutive 1's.

A zero followed by $r - 1$ ones occurs exactly once. But this is already accounted for by the run of r ones, which is preceded by a zero. Similarly, $r - 1$ ones followed by a zero occurs once, and is accounted for by the fact that the run of r ones must be followed by a zero. Thus, there is no true run of $r - 1$ ones.

Suppose $0 < k < r - 1$. To find the number of runs of ones of length k, consider r consecutive terms beginning with zero, then k

ones, then a zero, and the remaining $r - k - 2$ terms arbitrary. This occurs 2^{r-k-2} ways, since each way of completing the r terms occurs exactly once.

Analogous reasoning holds for the number of runs of 0's of length k, $0 < k < r - 1$. There is no run of r consecutive 0's (since this would "stop" the shift register). Yet "1 followed by $r - 1$ zeros" must occur, so there is a run of $r - 1$ zeros.

The structure of a maximum-length shift register sequence has thus been completely determined, insofar as runs of 1's (called *blocks*) and runs of 0's (called *gaps*) are concerned:

If $0 < k < r - 1$, there are 2^{r-k-2} blocks and an equal number of gaps of length k. Also, there is one gap of length $r - 1$ and one block of length r.

In terms of the period p of the sequence ($p = 2^r - 1$), there are $(p + 1)/2$ runs, half of them blocks and half of them gaps. Of the blocks, half have length 1, one-fourth have length 2, one-eighth have length 3, etc. and likewise for the gaps. This continues until there is one block and one gap of length $r - 2$. Beyond these, there is a gap of length $r - 1$ and a block of length r.

These results are most conveniently summarized in the form of

> **Theorem 4.2.** The randomness property $R - 2$ holds for all maximum-length shift register sequences.

4.3. *The Abelian group*

Let $A_1 = \{a_1, a_2, a_3, \ldots\}$ be a maximum-length shift register sequence with period $p = 2^r - 1$. Let $A_2 = \{a_2, a_3, \ldots\}$,

$A_3 = \{a_3, a_4, \ldots\}, \ldots, A_p = \{a_p, a_{p+1}, \ldots\}$.

Also, let $A_0 = \{0, 0, 0, \ldots\}$.

> **Definition.** An *Abelian group* G is a set of elements which satisfy:
> 1. The sum of any two elements of G is still in G.
> 2. For any elements a, b, c in G, $a + b = b + a$, and $(a + b) + c = a + (b + c)$.
> 3. There is an element 0 which satisfies $a + 0 = a$ for every a in G.
> 4. Every element a has a negative $\bar{a}$ such that $a + \bar{a} = 0$.

Note: For a *multiplicative* Abelian group, replace *sums* by *products*, the 0 element by the 1 element, and the negative by the reciprocal.

> **Theorem 4.3.** The sequences $A_0, A_1, \ldots, A_p$ form an Abelian group with respect to the operation of termwise addition mod 2. ["Termwise addition mod 2" means that

if $B = \{b_1, b_2, b_3, \ldots\}$ and $C = \{c_1, c_2, c_3, \ldots\}$ then $B + C =$
$\{b_1 + c_1, b_2 + c_2, b_3 + c_3, \ldots\}$.]

Proof. For any i, $A_i + A_0 = A_i$ and $A_i + A_i = A_0$. Thus A_0 is
a 0 element for the group, and A_i is the negative of itself.

Let R be the linear recurrence relation satisfied by A_1. Then R
is also satisfied by $A_2, \ldots, A_p$, and trivially by A_0. Since R is *linear*,
it is also satisfied by $A_i + A_j$ whenever it is satisfied by both A_i and
A_j, where $A_i + A_j$ denotes the term-by-term sum. Thus $A_i + A_j$ is
determined by R from its first r terms. Whatever these r terms are,
they are the same as the first r terms of exactly one of the 2^r sequences
$A_0, A_1, \ldots, A_p$. Thus, the sum of two sequences is still in the group.

The commutative law $[a + b = b + a]$ and the associative law
$[a + (b + c) = (a + b) + c]$ hold for the sequences because they hold for
ordinary addition mod 2.

4.4. *Randomness postulate R − 3*

Let $\{b_n\}$ be the sequence which results from $\{a_n\}$ by $b_n = 1 - 2 a_n$;
that is, 0's are replaced by 1's and 1's by -1's. This can also be
written: $b_n = e^{i\pi a_n}$.

Let $A_0, A_1, \ldots, A_p$ be defined as in 4.3. Replace 0's by 1's and
1's by -1's to get the sequences $B_0, B_1, \ldots, B_p$. Then addition mod
2 of the A_i's is the same as multiplication of the B_i's. This is clear
from the addition and multiplication tables:

(mod 2)

+	1	0		×	−1	1
1	0	1		−1	1	−1
0	1	0		1	−1	1 .

$$(1)$$

Since $A_i + A_j = A_k$, it follows that $B_i B_j = B_k$, where the product
$B_i B_j$ is taken term-by-term. Also, unless $B_k = B_0 = \{1, 1, 1, \ldots\}$, which
happens only if $i = j$, B_k contains $(p - 1)/2$ 1's and $(p + 1)/2 - 1$'s.

The auto-correlation of $\{b_n\}$ is

$$C(\tau) = \frac{1}{p} \sum_{n=1}^{p} b_n b_{n+\tau} = \begin{cases} 1 & \text{if } \tau = 0 \\ -1/p & \text{if } 0 < \tau < p, \end{cases} \quad (2)$$

because $\{b_n b_{n+\tau}\}$ is a sequence of the type $B_i \cdot B_j$, which in turn is of
the type B_k, and the excess of $+1$'s to -1's is p for B_0, and -1
otherwise. Thus,

Theorem 4.4. Every maximum-length shift register sequence has the randomness property $R - 3$.

*4.5.   *The orthogonality of group characters*

It is possible to relate the auto-correlation property just described to the theory of group characters.

The Abelian group $B_0, B_1, \ldots, B_p$ consists entirely of elements of order 2 (i.e., $B_i^2 = B_0$ for all i). It is known that such a group is a direct product of r groups, each with 2 elements, and the character table of the big group is the direct product of the character tables of the r small groups. Thus the character table of the group $G = \{B_0, B_1, \ldots, B_p\}$ consists entirely of 1's and -1's. If the *principal character*, consisting entirely of 1's, is adjoined horizontally to the sequences $B_0, B_1, \ldots, B_p$ written vertically, this forms the *character table* of the group G. From this point of view, Theorem 4.4 is simply a restatement of the Orthogonality Theorem for group characters.

For example, if the basic shift register sequence is $-1\ -1\ -1$ $1\ -1\ 1\ 1$, the character table is

$$
\begin{array}{c|cccccccc}
1 & 1 & 1 & 1 & 1 & 1 & 1 & 1 \\
\hline
1 & -1 & -1 & -1 & 1 & -1 & 1 & 1 \\
1 & -1 & -1 & 1 & -1 & 1 & 1 & -1 \\
1 & -1 & 1 & -1 & 1 & 1 & -1 & -1 \\
1 & 1 & -1 & 1 & 1 & -1 & -1 & -1 \\
1 & -1 & 1 & 1 & -1 & -1 & -1 & 1 \\
1 & 1 & 1 & -1 & -1 & -1 & 1 & -1 \\
1 & 1 & -1 & -1 & -1 & 1 & -1 & 1.
\end{array}
\tag{3}
$$

Any two rows (or columns) are readily seen to be orthogonal.

It is also possible to prove the following interesting converse of Theorem 4.3.

> *Theorem 4.5.*   Let $S_1 = \{s_1, s_2, s_3, \ldots\}$ be any sequence mod 2 with period p, for which $S_1 + S_i = S_k$ where $S_i = \{s_i, s_{i+1}, s_{i+2}, \ldots\}$, and either $S_k = \{s_k, s_{k+1}, s_{k+2}, \ldots\}$ or $S_k = \{0, 0, 0, \ldots\}$.
> Then $p = 2^r - 1$ for some integer r, and S_1 is in fact a maximum-length shift register sequence.

Proof. $S_0, S_1, \ldots, S_p$ form an Abelian group under term-by-term addition mod 2, in which every element has order 2. Such a group is known to be a direct sum of r groups of order 2, so that $p + 1 = 2^r$. Moreover, any r non-zero elements form a basis for this group. In particular, $S_{r+1} = \sum_{i=1}^{r} c_i S_i$. Viewed component-wise, this is a recursion formula of degree r for the sequence S_1. Thus S_1 is a shift register sequence of maximum length $p = 2^r - 1$.

*4.6.   *Legendre sequences*[1]

Let p be an odd prime number. The *Legendre symbol* (n/p) is defined as follows:

$$\left(\frac{n}{p}\right) = \begin{cases} 1 & \text{if there is an integer } x \text{ for which } x^2 \equiv n \pmod p \\ -1 & \text{otherwise}. \end{cases} \quad (4)$$

For example, if $p = 7$, the perfect squares mod 7 are 0, 1, 4, 2. Hence

$$\left(\frac{0}{7}\right) = 1, \quad \left(\frac{1}{7}\right) = 1, \quad \left(\frac{2}{7}\right) = 1, \quad \left(\frac{3}{7}\right) = -1,$$

$$\left(\frac{4}{7}\right) = 1, \quad \left(\frac{5}{7}\right) = -1, \quad \left(\frac{6}{7}\right) = -1.$$

O. Perron [28] observes that the sequence $(0/p), (1/p), (2/p) \cdots$ has the property $R - 3$, provided p is a prime of the form $4n - 1$. Thus the sequence $1\ 1\ 1\ -1\ 1\ -1\ -1$ satisfies $R - 3$. Except for these "Legendre sequences" for $p = 3$ and $p = 7$, the sequences he gets are *not* maximum-length shift register sequences. (*Note*: Interchanging $+1$'s and -1's is an inessential distinction, in that it does not affect the auto-correlation; and two sequences which differ only in this regard will not usually be considered different.)

In general, $R - 2$ does not hold for Legendre sequences. Perron actually proves that there are $(p + 1)/2$ runs, including $(p + 1)/4$ blocks and $(p + 1)/4$ gaps; and that $(p + 1)/4$ runs are of length 1. But more is not usually the case. Thus, example B of Section 1 is the Legendre sequence for $p = 11$, and it fails to satisfy $R - 2$. More strikingly, the Legendre sequence of length 31 does not satisfy $R - 2$ either. Perron raised the question (though not in these words): "What are all the sequences which satisfy $R - 1$ and $R - 3$?" J. B. Kelly [21] believed it possible that the Legendre sequences are the *only* sequences of prime length with these properties. However, for every Mersenne prime period $p > 7$ there are shift register sequences with these properties which are *not* Legendre sequences.

Traditionally, the Legendre symbol (a/p) is either undefined when $a = 0$, or else defined as $(0/p) = 0$. However, to get Perron's examples,

---

[1] The discussion in 4.6 was written before I became aware of the literature on "difference sets." What is called "Perron's Problem" is more generally known as the question of the existence of "Hadamard difference sets." See, for example, [18].

we must use either $+1$ or -1 in this position. Either choice yields the desired correlation properties.

In the next section, Perron's question is, in essence, answered. In addition, the more general question of sequences which satisfy $R-3$ is treated.

4.7. *Summary*

Maximum-length shift register sequences satisfy all three of the Randomness Postulates of Section 1. There is another family of sequences, called Legendre sequences, which satisfy $R-1$ and $R-3$, but not, in general, $R-2$. Actually, the three randomness postulates characterize the maximum-length shift register sequences, which are thus the true "pseudo-noise sequences." The sequences which satisfy $R-1$ and $R-3$ correspond to a topic in the recent literature on combinatorial analysis called "Hadamard difference sets." In addition to the shift register sequences and the Legendre sequences, there are other examples as well. Some of these are discussed in [19].

5.   CYCLOTOMIC POLYNOMIALS

5.1.   *Algebraic background*

Let $f(x)$ be a polynomial mod 2 of degree $r > 1$, with $f(0) = 1$ (i.e., the constant term is 1). Then

$$\frac{1}{f(x)} = \sum_{n=0}^{\infty} a_n x^n , \tag{1}$$

where the sequence $\{a_n\}$ can be generated by a shift register. By Theorem 2.1, the sequence $\{a_n\}$ is periodic. Let its period be p. Then $f(x)$ divides $1 - x^p$, by Theorem 2.3.

Since $f(x)$ divides $1 - x^p$, every root of $f(x)$ is also a root of $1 - x^p$. The roots of $1 - x^p = 0$ (i.e., of $x^p - 1$) are called the pth *roots of unity*. Geometrically, they are the vertices of a regular $p-$gon with center at the origin and one vertex at the point 1.

Thus, every shift register polynomial is simply the equation satisfied by certain pth roots of unity. (The other polynomials mod 2 look like $x^p \cdot f(x)$, and have the root $x = 0$ in addition to roots which are roots of unity.)

The periods of *irreducible* polynomials are always odd, because they divide $2^r - 1$. (In fact, the periods of "square-free" polynomials— polynomials without repeated factors—are odd, because they are the L. C. M. of the periods of the individual irreducible factors.) Conversely, every odd number divides a number of the type $2^r - 1$. Thus, for the most part, if attention is restricted to $(2^r - 1)$st roots of unity, all the interesting cases will be covered.

5.2. *Cyclotomic polynomials*

In the complex plane, the pth roots of unity can be written $z = e^{2n\pi i/p}$, with $n = 1, 2, \ldots, p$. If n/p is an irreducible fraction, the corresponding pth root is called *primitive*. By 3.2, there are $\phi(p)$ *primitive* pth roots of unity. By reducing n/p to lowest terms, every pth root is a *primitive* qth root, where q is some factor of p.

When working mod 0 (the ordinary complex numbers), all the primitive pth roots of unity satisfy the same irreducible equation [39]. For example, the primitive cube roots, $e^{2\pi i/3}$ and $e^{4\pi i/3}$ satisfy $x^2 + x + 1 = 0$. This equation is called the *cyclotomic polynomial* of order p, and has degree $\phi(p)$.

Explicitly, it is known [39] that this polynomial is given by

$$C_p(x) = \prod_{d|p} (x^d - 1)^{\mu(p/d)} \tag{2}$$

where the product is extended over all divisors d of p, and the exponent is the Möbius μ-function of 3.2.

5.3. *Factorization mod 2*

The problem which now arises is: How does $C_p(x)$ factor modulo 2? For it is the *factors* of $C_p(x)$ which are the irreducible polynomials of period p, that is, they divide $1 - x^p$ but *do not* divide $1 - x^d$ for any d smaller than p, since these factors are removed in equation (2).

In particular, the factors of $C_{2^r-1}(x)$ are the characteristic polynomials for maximum-length shift register sequences. These polynomials are known each to have degree r. The product of them all, namely $C_{2^r-1}(x)$, has degree $\phi(2^r - 1)$. Hence there are $\phi(2^r - 1)/r$ maximum exponent polynomials of degree r, whence the formula

$$\lambda(r) = \phi(2^r - 1)/r \tag{3}$$

of 3.4.

*5.4. *The Ψ_2 formula*

The irreducible polynomials of degree r all divide $1 - x^{2^r-1}$. Their periods divide $2^r - 1$ (while dividing no smaller number $2^s - 1$). If t is a divisor of $2^r - 1$ which divides no smaller $2^s - 1$, then analogous to the above, there are $\phi(t)/r$ irreducible polynomials of exponent t. Thus

$$\Psi_2(r) = \sum_{\substack{t|(2^r-1) \\ t \nmid (2^s-1)}} \frac{\phi(t)}{r},$$

or

$$\Psi_2(r) = \frac{1}{r} \sum_{d \mid k} (2^r - 1)\mu\left(\frac{r}{d}\right),$$ (4)

using

$$\sum_{t \mid 2^r} \phi(t) = 2^r - 1, \qquad \text{from 3.4.}$$

Using

$$\sum_{d \mid r} \mu\left(\frac{r}{d}\right) = \sum_{d \mid r} \mu(d) = 0 \qquad \text{for } r > 1,$$

$$\Psi_2(r) = \frac{1}{r} \sum_{d \mid r} 2^r \mu\left(\frac{r}{d}\right),$$ (5)

in agreement with 3.4.

(The standard proof in the literature is more direct in that it does not involve (3), but makes use of the "Möbius inversion formula," which need not be introduced.)

5.5.   *The cyclotomic cosets*

The number of factors of $C_{2^r-1}(x)$ has been determined. However, the question of how the roots $e^{2n\pi i/p}$ are separated into sets of roots for the maximum period polynomials has not yet been answered.

The integers n from 1 to $p-1$ with no factors in common with p form a multiplicative Abelian group (see 4.3), with respect to the operation: multiplication modulo p. The numbers $1, 2, 4, \ldots, 2^{r-1}$ form a *subgroup* with r elements. This subgroup, when multiplied by any other element of the group, yields a *coset*.

For example, if $p = 2^5 - 1 = 31$, the multiplicative group consists of the integers from 1 to 30. The decomposition into cosets is

$$
\begin{array}{llllll}
C_0: & 1 & 2 & 4 & 8 & 16 \\
C_1: & 3 & 6 & 12 & 24 & 17 \\
C_2: & 9 & 18 & 5 & 10 & 20 \\
C_3: & 27 & 23 & 15 & 30 & 29 \\
C_4: & 19 & 7 & 14 & 28 & 25 \\
C_5: & 26 & 21 & 11 & 22 & 13 \, .
\end{array}
$$ (6)

The roots of any factors of $C_p(x)$ mod 2 are the numbers ϵ^n, where n ranges through a fixed *coset* in the multiplicative group mod p, and ϵ is a primitive pth root of unity.

Gauss [10], who was the first to study cyclotomic factorization, referred to the cosets as the *periods* of the cyclotomic equation. However, to avoid confusion with the periodic structure of everything else, the term *cosets* will be used instead throughout this work.

5.6. *A necessary condition for* $R - 3$

The cyclotomic cosets will reappear in 5.11. Meanwhile, it is necessary to study the randomness axiom $R - 3$, in preparation for showing its connection with the cosets.

$R - 3$ states: The sequence $A = \{a_n\}$ of $+1$'s and -1's with period p satisfies

$$pC(\tau) = \sum_{n=1}^{p} a_n a_{n+\tau} = \begin{cases} p & \text{if } \tau = 0 \\ K & \text{if } 0 < \tau < p \end{cases}. \tag{7}$$

That is, the auto-correlation of A is two valued. (If $p > 1$ then $K < p$, because $a_1, a_2, \ldots, a_p$ cannot all be identical.)

Let A contain $a + 1$'s and $b - 1$'s. First,

$$p = a + b. \tag{8}$$

Second,

$$\sum_{n=1}^{p} a_n = a - b. \tag{9}$$

Third,

$$\sum_{\tau=1}^{p} pC(\tau) = \sum_{\tau=1}^{p} \sum_{n=1}^{p} a_n a_{n+\tau} = \sum_{n=1}^{p} a_n \sum_{\tau=1}^{p} a_{n+\tau} = (a - b)^2. \tag{10}$$

On the other hand,

$$\sum_{\tau=1}^{p} pC(\tau) = pC(0) + \sum_{\tau=1}^{p-1} pC(\tau) = p + K(p - 1). \tag{11}$$

From (10) and (11),

Theorem 5.1. If A satisfies $R - 3$, then

$$K = \frac{(a - b)^2 - 1}{p - 1}. \tag{12}$$

This is a *necessary* condition which the out-of-phase value K must satisfy.

Note that if $a - b = 1$, then $K = -1$, and conversely. Thus $R - 1$ and $R - 3$ imply that

$$C(\tau) = \begin{cases} 1 & \text{if } \tau = 0 \\ -1/p & \text{if } 0 < \tau < p \end{cases}, \tag{13}$$

as satisfied by both the Perron sequences and the maximum-length shift register sequences.

If $a = b$, $K = -p/(p - 1)$, which is *never* an integer, for $p > 2$. That is, $R - 3$ is never satisfied with the same number of $+1$'s as

− 1's in the sequence (unless $p = 2$, which is the exception that proves the rule).

In (12), the roles of a and b are interchangeable, since interchanging + 1's and − 1's does not affect the auto-correlation function.

If $a = 1$, then $b = p − 1$. In this case,

$$K = \frac{(p - 2)^2 - p}{p - 1} = \frac{p^2 - 5p + 4}{p - 1} = p - 4 \,. \qquad (14)$$

Such sequences always satisfy $R - 3$. In fact, a sequence of all − 1's except for one + 1 is a pulse. By (14), the auto-correlation of a pulse is 1 in phase, and $(p - 4)/p$ out of phase.

If $p = 4$, this yields a sequence with 0 as the out-of-phase correlation level, namely 1 −1 −1 −1. In general, from (12), $K = 0$ implies

$$p = (a + b) = (a - b)^2 \,. \qquad (15)$$

The condition $K = + 1$ leads to $p - 1 = (a - b)^2 - p$, or

$$2p - 1 = (a - b)^2 \,. \qquad (16)$$

An example of this situation occurs with $a = 9$, $b = 4$, so that $p = 4 + 9 = 13$. Then

$$2 \cdot 13 - 1 = 25 = (9 - 4)^2 \,.$$

Such a sequence of length 13 is 1 1 1 1 1 −1 1 1 1 −1 −1 1 −1.

5.7. *Another necessary condition*

Another *necessary* condition for a sequence to satisfy $R - 3$ is that K and p have the same parity; that is, $K + p$ must be *even*.

Theorem 5.2. If A satisfies $R - 3$, then $K \equiv p$ (mod 2).

Proof. $p = a + b$. In forming the out-of-phase correlation, suppose $\sum a_n a_{n+\tau}$ contains 1·1 as a term y times, and contains $−1 \cdot −1$ z times. Then $1 \cdot - 1$ occurs $a - y$ times, since the first factor is + 1 exactly a times. But $1 \cdot - 1$ occurs $b - z$ times, because the *second* factor is − 1 exactly b times. Thus $a - y = b - z$. Moreover, $- 1 \cdot 1$ occurs $a - y = b - z$ times. Finally,

$$K = \sum_{n=1}^{p} a_n a_{n+\tau} = y - (a - y) - (b - z) + z = 2(y + z) - (a - b) \,. \quad (17)$$

Reducing modulo 2, $K \equiv a + b \equiv p$ (mod 2).

Theorem 5.1 does *not* imply Theorem 5.2. For example, with $a = 6$ and $b = 3$, (12) becomes

$$K = \frac{(6 - 3)^2 - 9}{9 - 1} = 0 \,,$$

which is an integer. But $p = 9$ is odd, while $K = 0$ is even, so that by Theorem 5.2, there is no such sequence satisfying $R - 3$.

Moreover, it is *not* true that Theorems 5.1 and 5.2 together are *sufficient* for the existence of a sequence satisfying $R - 3$. Other conditions must also be satisfied. The nature of such conditions is discussed in 5.12.

*5.8. *The case of non-binary sequences*

Theorem 5.1 can be generalized to a wider class of sequences.

> **Theorem 5.3.** Let $S = \{a_n\}$ be a sequence of real numbers, with period p, satisfying
>
> $$\sum_{n=1}^{p} a_n a_{n+\tau} = \begin{cases} p & \text{if } \tau = 0 \\ K & \text{if } 0 < \tau < p \end{cases}.$$

Then

$$K = \frac{(A - B)^2 - p}{p - 1}, \tag{18}$$

where A is the sum of the *positive* terms, and B the sum of the *negative* terms of S.

$$\left(\text{Thus, } \sum_{n=1}^{p} a_n = A - B.\right)$$

The proof of this theorem is identical to that of Theorem 5.1.

This is the only general result about non-binary sequences which will be included here. One specific result is that any sequence of real numbers with period 3 has a two-level auto-correlation, namely

$$a_1^2 + a_2^2 + a_3^2 \qquad \text{in phase,}$$
$$a_1 a_2 + a_2 a_3 + a_3 a_1 \qquad \text{out of phase}.$$

5.9. *Decimation of $R - 3$ sequences*

Suppose again that $A = \{a_n\}$ is a binary sequence, with the $R - 3$ property:

$$\sum_{n=1}^{p} a_n a_{n+\tau} = \begin{cases} p & \text{if } \tau = 0 \\ K & \text{if } 0 < \tau < p \end{cases}.$$

Suppose q is any integer relatively prime to p (i.e., q/p is an irreducible fraction). Then the numbers $q, 2q, 3q, \ldots, pq$ are the same mod p as $1, 2, 3, \ldots, p$, except for the order in which they occur. (This is the "Rearrangement Theorem" of Group Theory.) Thus, $a_q, a_{2q}, \ldots, a_{pq}$ is simply a rearrangement (or permutation) of $a_1, a_2, \ldots, a_p$, and the *sum* of the two is the same.

Then

$$\sum_{n=1}^{p} a_{qn} a_{qn+\tau} = \sum_{n=1}^{p} a_n a_{n+\tau},$$

or finally,

$$\sum_{n=1}^{p} a_{qn} a_{qn+q\tau} = \begin{Bmatrix} p & \text{if } \tau = 0 \\ K & \text{if } 0 < \tau < p \end{Bmatrix}. \tag{19}$$

But if $0 < \tau < p$ then reducing $q\tau$ mod p, $0 < q\tau < p$. That is,

> *Theorem 5.4.* If $\{a_n\}$ satisfies $R - 3$, so does $\{a_{qn}\}$, pro-
> vided q is prime to the period p.

The equation here is

$$\sum_{n=1}^{p} a_{qn} a_{qn+\tau} = \begin{Bmatrix} p & \text{if } \tau = 0 \\ K & \text{if } 0 < \tau < p \end{Bmatrix}. \tag{20}$$

This remarkable fact can be rephrased as follows: If a sequence satisfies $R - 3$, then every "proper decimation" of it does also. *Decimation* means selecting every q^{th} element from the sequence. *Proper* means that q is to be relatively prime to the period p.

Example. The sequence 1 1 1 1 1 −1 1 1 1 −1 −1 1 −1 has been mentioned as satisfying $R - 3$. Selecting every second term, the sequence 1 1 −1 1 −1 1 1 1 1 1 −1 −1 results. This is essentially a new sequence (i.e., not a cyclic permutation of the previous one), and by Theorem 5.4, it still has the "random auto-correlation" property $R - 3$.

5.10. *Subgroup and cosets*

There are $\phi(p)$ numbers q, $0 < q < p$, which are relatively prime to p, and as already observed, they form an Abelian group G with respect to multiplication modulo p.

Let $A_1 = \{a_n\}$ be a binary sequence of period p which satisfies $R - 3$. By Theorem 5.4, $A_q = \{a_{qn}\}$ also satisfies $R - 3$, for every q in the group. Let C_0 be the set of q's in G for which A_q is simply a phase shift of A_1 (i.e., not essentially different). Then C_0 is a *subgroup* of G, for if $\{a_{qn}\}$ and $\{a_{rn}\}$ are simple phase shifts of $\{a_n\}$, then $\{a_{qrn}\} = \{a_{q(rn)}\}$, and a phase shift of a phase shift is still just a phase shift.

Designate the *cosets* of C_0 by $C_1, C_2, \ldots, C_m$.

> *Theorem 5.5.* The sequences $\{a_{qn}\}$ and $\{a_{rn}\}$ differ by no
> more than a phase shift if and only if q and r belong to
> the same coset of C_0.

Proof. If the numbers q and r belong to the same coset of C_0 then they differ only by a factor from C_0. Let that factor be e: $q = er$.

Since e is in C_0, $\{a_{rn}\}$ is a simple phase shift of $\{a_n\}$. Likewise, $\{a_{qn}\} = \{a_{e(rn)}\}$ and $\{a_{qn}\}$ is a simple phase shift of $\{a_{rn}\}$.

Conversely, if $\{a_{qn}\}$ is a phase shift of $\{a_{rn}\}$, let r^{-1} be the inverse of r in G, and let $qr^{-1} = e$. Then $q = er$. Since $\{a_{qn}\} = \{a_{e(rn)}\}$ is simply a phase shift of $\{a_{rn}\}$, e must belong to C_0. Then q and r differ only by a factor in C_0, and therefore are in the same *coset* of C_0.

Example. Let $\{a_n\} = 1\ 1\ 1\ -1\ 1\ -1\ -1$. Thus $p = 7$. The permissible values of q are $1, 2, 3, 4, 5, 6$. For $q = 1, 2, 4$, $\{a_{qn}\} = \{a_n\}$, while for $q = 3, 5, 6$, $\{a_{qn}\}$ is $\{a_n\}$ running *backwards*.

Thus, $C_0 = \{1, 2, 4\}$, and $C_1 = \{3, 5, 6\}$.

5.11. *The coset decomposition*

If $p = 2^r - 1$ and $C_0 = \{1, 2, 4, \ldots, 2^{r-1}\}$ then the coset structure of Theorem 5.5 is the same as the cyclotomic cosets of 5.5. It may be noted that this is always the case for maximum-length shift register sequences. But first,

> *Theorem 5.6.* If $A = \{a_n\}$ satisfies $R - 3$, and C_0 is the subgroup of elements q such that $\{a_{qn}\}$ is simply a phase shift of $\{a_n\}$, then there is a phase shift $B = \{b_n\}$ of A which is left completely invariant by C_0.

Proof. Suppose C_0 has a primitive element q. That is, $C_0 = \{1, q, q^2, \ldots, q^u\}$, where powers are reduced modulo p. Suppose $\{a_{qn}\} = \{a_{n+a}\}$. If $\alpha = 0$, there is no problem. Otherwise, it is always possible to pick $\{b_n\} = \{a_{n+\tau}\}$ in such a way that $\{b_{qn}\} = \{b_n\}$. Then also $\{b_{q^2n}\} = \{b_{q(qn)}\} = \{b_{qn}\} = \{b_n\}$, and similarly for all the other elements q of C_0. Thus C_0 leaves $B = \{a_{n+\tau}\}$ completely invariant.

Finally, if C_0 has two generators, a double application of the above process will work. (The case of more than two generators is of course treated similarly.)

> *Theorem 5.7.* A sequence A which satisfies $R - 3$, with period p, can be broken up into *cosets* of $+1$'s and of -1's.

Proof. By Theorem 5.6, it is no loss of generality to assume that C_0 leaves A completely invariant. Let $A = \{a_0, a_1, a_2, \ldots, a_p\}$. If q is in C_0, then

$$a_1 = a_q = a_{q^2} = \cdots,$$
$$a_2 = a_{2q} = a_{2q^2} = \cdots,$$
$$a_3 = a_{3q} = a_{3q^2} = \cdots, \text{ etc}.$$

That is, if the term a_1 is $+1$, so are the terms $a_q, a_{q^2}, \ldots$ and in fact,

$a_c = 1$ for every c in C_0. Similarly, if α and β are both in the same coset C_j, then $a_\alpha = a_\beta$.

Example. The sequence $\{a_n\} = 1\,1\,1\,-1\,1\,-1\,-1$ is left completely invariant by $C_0 = \{1, 2, 4\}$ and is inverted by $C_1 = \{3, 5, 6\}$. The a_0 element is $+1$. Thus $a_0 = 1$,

$$a_1 = a_2 = a_4 = 1, \text{ and } a_3 = a_5 = a_6 = -1.$$

5.12. *Superposition of cosets*

If $\{a_n\}$ is a maximum-length shift register sequence, with period $p = 2^r - 1$, then $C_0 = \{1, 2, 4, \ldots, 2^{r-1}\}$, because this gives the only decomposition into cosets where there are $\lambda_2(r) = (1/r)\phi(2^r - 1)$ cosets (it being evident that $\{a_n\}$ is transformed into the other maximum-length shift register sequences by the elements of the multiplicative group).

From this, it is possible to construct shift register sequences by *superposition* of the cosets.

For example, to find the shift register sequences which have $r = 5$ and $p = 31$, the multiplicative group (mod 31) comes into play. Using $C_0 = (1, 2, 4, 8, 16)$ as a subgroup, the cosets are

$$
\begin{array}{lrrrrr}
C_0: & 1 & 2 & 4 & 8 & 16 \\
C_1: & 3 & 6 & 12 & 24 & 17 \\
C_2: & 9 & 18 & 5 & 10 & 20 \\
C_3: & 27 & 23 & 15 & 30 & 29 \\
C_4: & 19 & 7 & 14 & 28 & 25 \\
C_5: & 26 & 21 & 11 & 22 & 13 .
\end{array}
\tag{21}
$$

The next question which must be answered is how to assign the values $+1$ and -1 to the various cosets. This is taken up in detail in Section 6. For the present, suffice it to say that the following six assignments are valid:

	C_0	C_1	C_2	C_3	C_4	C_5
S_1	-1	1	1	-1	-1	1
S_2	1	-1	1	1	-1	-1
S_3	-1	1	-1	1	1	-1
S_4	-1	-1	1	-1	1	1
S_5	1	-1	-1	1	-1	1
S_6	1	1	-1	-1	1	-1.

$$\tag{22}$$

Assigning 0 the value 1 for all cases, the six shift register sequences are:

Table III-4.

	C_0	C_1	C_2	C_3	C_4	C_5	S_1	S_2	S_3	S_4	S_5	S_6
0							1	1	1	1	1	1
1	1						—	1	—	—	1	1
2	1						—	1	—	—	1	1
3		1					1	—	1	—	—	1
4	1						—	1	—	—	1	1
5			1				1	1	—	1	—	—
6		1					1	—	1	—	—	1
7				1			—	—	1	1	—	1
8	1						—	1	—	—	1	1
9			1				1	1	—	1	—	—
10			1				1	1	—	1	—	—
11						1	1	—	—	1	1	—
12		1					1	—	1	—	—	1
13						1	1	—	—	1	1	—
14				1			—	—	1	1	—	1
15				1			—	1	1	—	1	—
16	1						—	1	—	—	1	1
17		1					1	—	1	—	—	1
18			1				1	1	—	1	—	—
19				1			—	—	1	1	—	1
20			1				1	1	—	1	—	—
21						1	1	—	—	1	1	—
22						1	1	—	—	1	1	—
23				1			1	1	—	1	—	—
24		1					1	—	1	—	—	1
25				1			—	1	1	—	1	—
26						1	—	1	—	—	1	1
27				1			—	1	1	—	1	—
28				1			—	—	1	1	—	1
29			1				1	1	—	1	—	—
30			1				—	1	—	1	—	—

The slightly more complicated case, when p is composite, as well as the general problem of determining the array (22), are considered in Section 6.

It should be mentioned that the Legendre sequences can also be gotten by the method (23) just employed. In fact, 0, C_0, C_2, and C_4 are the *quadratic residues* mod 31, while C_1, C_3, C_5 are the *quadratic non-residues*. This fact also makes it possible to mechanize the Legendre sequences as mod 2 sums of shift register sequences.

*5.13. *The Bruck-Ryser method*

In 5.6 and 5.7, necessary conditions were obtained for a sequence to satisfy $R - 3$. To obtain additional restrictions on p and K, the following method may be employed: Let $A = \{a_n\}$ satisfy

$$\sum_{n=1}^{p} a_n a_{n+\tau} = \begin{Bmatrix} p & \text{if } \tau = 0 \\ K & \text{if } 0 < \tau < p \end{Bmatrix}.$$

Consider the matrix

$$M = \begin{pmatrix} a_1 & a_2 & \cdots & a_p \\ a_2 & a_3 & \cdots & a_1 \\ a_3 & a_4 & \cdots & a_2 \\ - & - & - & - \\ a_p & a_1 & \cdots & a_{p-1} \end{pmatrix}. \tag{24}$$

This matrix is symmetric; i.e., $M = M^t$, the *transpose matrix* of M. Moreover, simple matrix multiplication shows

$$M^2 = MM^t = \begin{pmatrix} p & K & K & \cdots & K \\ K & p & K & \cdots & K \\ K & K & p & \cdots & K \\ & & & \ddots & \\ - & - & - & - & - \\ K & K & K & \cdots & p \end{pmatrix}. \tag{25}$$

A completely analogous situation, which arises in the determination of finite projective planes, has been studied by Bruck and Ryser [2]. Their methods involve the Hilbert norm-residue symbol, and the evaluation of Hecke-Minkowski invariants. The particulars are beyond the scope of the present work.

5.14. *Summary*

The maximum index polynomials of degree r are the factors, mod 2, of the cyclotomic polynomial of order $2^r - 1$. The roots of this polynomial are the "primitive $(2^r - 1)$st roots of unity." The other irreducible polynomials of degree r have non-primitive $(2^r - 1)$st roots of unity as their roots. The formulas of Section 3 can be derived from these facts.

The primitive $(2^r - 1)$st roots of unity form a multiplicative group. When the cyclotomic polynomial which they satisfy is factored, the roots are associated with the various factors in accordance with certain *cosets* of the group. It is shown how these are computed.

The randomness postulate $R - 3$ is studied independently of $R - 1$ and $R - 2$. Two important necessary conditions are found in order that $R - 3$ may hold. Then, it is seen that sequences which satisfy $R - 3$ can be decomposed into a coset structure, very much like the cyclotomic cosets. This is illustrated in some special cases, and will be more widely applied in Section 6.

(For the algebraically sophisticated reader, it is worth mentioning that the transformations $\{a_i\} \rightarrow \{a_{qi}\}$ with q in C_0 are *automorphisms* of the field containing the roots of the characteristic polynomial for the $\{a_i\}$ sequence. Since we know that $q = 1, 2, 4, \ldots, 2^{n-1}$ all leave $\{a_i\}$ fixed, we have examples of n such automorphisms. However, the Fundamental Theorem of Galois Theory asserts that there cannot be more than n automorphisms, when the field is generated by the roots of an irreducible polynomial of degree n. Hence, C_0 contains *only* $q = 1, 2, 4, \ldots, 2^{n-1}$; and the other cosets must yield the sequences corresponding to the other primitive polynomials.)

6. COMPUTATIONAL TECHNIQUES

6.1. *Introduction*

It has now been shown that pseudo-noise sequences, as defined in Section 1, are generated by shift registers with the proper feedback connections. It has also been shown that the connections are "proper" if a certain polynomial $f(x)$ of degree r (where r is the number of tubes in the shift register) is a factor of the cyclotomic polynomial $C_{2^r-1}(x)$, modulo 2. Specifically, $f(x)$ is obtained by

$$f(x) = 1 + \sum_{i \text{ in } F} x^i \qquad (1)$$

where the powers of x which occur are the numbers of the tubes which are used in the feedback. Thus, there is an immediate correspondence between the shift registers and the polynomials.

The problem of determining which feedbacks are appropriate reduces, then, to the listing of certain polynomials. It is to this problem that the present chapter is devoted.

6.2. *The two basic methods*

There are two essentially different points of view which one may adopt concerning the nature of the polynomials which correspond to maximum-length shift register sequences. On the one hand, they are the factors of $C_{2^r-1}(x)$ mod 2. On the other, they are the irreducible polynomials mod 2 of degree r, from which a few deletions have been made. In the case that $2^r - 1$ is prime, it has been pointed out (3.1) that no deletions are necessary.

A computational technique corresponds to each of these points of view:

To compute a table of irreducible polynomials, the best approach is the "sieve method," which is the same as the method used by the ancients for listing the prime numbers. In this *analytic* (i.e., "tearing down") approach to the problem, *all* the polynomials are listed, and then those which have non-trivial factors are crossed off.

To factor cyclotomic polynomials, some sort of *synthetic* (i.e., "building up") method should be employed. Such methods will directly involve the cyclotomic cosets of Section 5.

In the case that $2^r - 1$ is not prime, the sieve method must be supplemented by a *deletion* of certain extra polynomials. This deletion can be accomplished by an application of the synthetic techniques, on a much smaller scale than would be required to do all the work by the synthetic method.

6.3. *The sieve method*

If a polynomial $f(x)$ of degree r can be factored, then it has a factor of degree $\leq r/2$, since the product of two polynomials of degree $> r/2$ has degree $> r$. In fact, $f(x)$ has an *irreducible* factor of degree $\leq r/2$. For example, a polynomial of degree 19 is irreducible if (and only if) it is divisible by none of the 127 irreducible polynomials of degree ≤ 9.

Special devices can be used to test for divisibility by polynomials of very low degree. These depend on the well-known theorem: If $g(x)$ divides $f(x)$, then every root of $g(x)$ is also a root of $f(x)$. Thus, $f(x)$ is divisible by x if and only if $f(0) = 0$, since 0 is a root of $g(x) = x$. Since $f(0)$ is the *constant* term of $f(x)$, which is always 1 or 0 (mod 2), an *irreducible* polynomial must have the constant term 1.

Similarly, $g(x) = x + 1$ has the root 1: $g(1) = 1 + 1 = 0$ (mod 2). Thus, $f(x)$ is divisible by $x + 1$ if and only if $f(1) = 0$; i.e., if and only if $f(x)$ has an *even* number of terms.

Let α be a root of $g(x) = x^2 + x + 1$. Since $g(x)$ divides $1 - x^3$, $\alpha^3 = 1$. A polynomial $f(x)$ is divisible by $g(x)$ if and only if $f(\alpha) = 0$. Using $\alpha^3 = 1$, one may reduce the *exponents* of the terms in $f(x)$ modulo 3, and decide whether or not the *reduced* polynomial is divisible by $g(x)$. If $f(x)$ is already known to have no *first degree* factors (x and $x + 1$ are first degree), it suffices to observe whether or not $f(x)$ reduces to $x^2 + x + 1$. For example, $f(x) = x^7 + x^5 + x^4 + x + 1$ reduces to $x + x^2 + x + x + 1 = x^2 + x + 1$, so that $f(x)$ *is* divisible by $x^2 + x + 1$.

Since both irreducible polynomials of degree 3 ($x^3 + x + 1$ and $x^3 + x^2 + 1$) divide $1 - x^7$, it is possible to test a polynomial $f(x)$ for divisibility by either one of these in a single operation. Reduce the exponents of the terms in $f(x)$ modulo 7, and see if the result is divisible by either 3rd degree polynomial. Explicitly, the reduced polynomial is one of the following, if $f(x)$ has a 3rd degree factor (and is *not* divisible by x or $x + 1$):

(1) $x^3 + x + 1$,   (9) $x^5 + x + 1$,

(2) $x^3 + x^2 + 1$,   (10) $x^5 + x^4 + 1$,

(3) $x^4 + x^2 + x$,   (11) $x^6 + x^2 + x$,

(4) $x^4 + x^3 + x$,   (12) $x^6 + x^5 + x$,

(5) $x^5 + x^3 + x^2$,   (13) $x^6 + x^2 + 1$,   (2)

(6) $x^5 + x^4 + x^2$,   (14) $x^6 + x^4 + 1$,

(7) $x^6 + x^4 + x^3$,   (15) $x^6 + x^5 + x^4 + x^3 + x^2 + x + 1$.

(8) $x^6 + x^5 + x^3$,

Increasingly complex criteria can be furnished for divisibility by higher degree polynomials. However, there is a principle of diminishing returns by which the very complex criteria eliminate very few poly-nomials from the list. One other simple criterion exists:

A polynomial is reducible if all its exponents are even, for then it is a perfect square. E.g., $x^6 + x^2 + 1 = (x^3 + x + 1)^2$.

6.4. *Applying the sieve method*

The following illustration of the sieve method may be given: The list of irreducible 11th degree polynomials mod 2, which appears in Table III-5, was obtained in accordance with the following program:

1. Write down all the 11th degree polynomials with an *odd* number of terms, and ending in the constant term *1*. There are $2^9 = 512$ of these.

2. Cross out the polynomials which can be obtained as the product of an irreducible 5th degree times an irreducible 6th degree polynomial. There are $6 \times 9 = 54$ of these, leaving 458 on the list.

3. Cross out the polynomials which can be obtained as the product of an irreducible 4th degree times an irreducible 7th degree polynomial. There are $3 \times 18 = 54$ of these, leaving 404 on the list.

4. Reduce the exponents of the polynomials mod 3. Cross off those which then reduce to $x^2 + x + 1$. This eliminates 128 polynomials, leaving 276.

Table III-5. Irreducible polynomials modulo 2, through degree 11, with their periods.

Degree (bold) & polynomial*	Period	Degree (bold) & polynomial	Period	Degree (bold) & polynomial	Period
1		211	127	613	85
2	—	217	127	615	255
3	1	221	127	637	51
2		235	127	643	85
7	3	247	127	651	255
		253	127	661	51
3		271	127	675	85
13	7	277	127	703	255
15	7	301	127	717	255
4		313	127	727	17
23	15	323	127	735	85
31	15	325	127	747	255
37	5	345	127	763	51
		357	127	765	255
5		361	127	771	85
45	31	367	127		
51	31	375	127	**9**	
57	31			1003	73
67	31	**8**		1021	511
73	31	433	51	1027	73
75	31	435	255	1033	511
		453	255	1041	511
6		455	255	1055	511
103	63	471	17	1063	511
111	9	477	85	1113	73
127	21	515	255	1137	511
133	63	537	255	1145	73
141	63	543	255	1151	511
147	63	545	255	1157	511
155	63	551	255	1167	511
163	63	561	255	1175	511
165	21	567	85	1207	511
7		573	85	1225	511
203	127	607	255	1231	73

* If $f(x) = \sum_{i=0}^{n} c_i x^i$, the table entry is $\sum_{i=0}^{n} c_i 2^i$ written to the base 8. Thus $x^5 + x^3 + x^2 + x + 1$ becomes binary 101, 111 which is octal "57."

Table III-5 (Cont'd).

Degree (bold) & polynomial	Period	Degree (bold) & polynomial	Period	Degree (bold) & polynomial	Period
1243	511	1773	511	2547	341
1245	511			2553	1023
1257	511	10		2605	1023
1267	511	2011	1023	2617	1023
1275	511	2017	341	2627	1023
1317	511	2033	1023	2633	341
1321	511	2035	341	2641	1023
1333	511	2047	1023	2653	341
1365	511	2055	1023	2671	341
1371	511	2065	93	2701	341
1401	73	2107	341	2707	1023
1423	511	2123	341	2745	1023
1425	511	2143	341	2767	1023
1437	511	2145	1023	2773	1023
1443	511	2157	1023	3023	1023
1461	511	2201	1023	3025	1023
1473	511	2213	1023	3043	33
1511	73	2231	341	3045	1023
1517	511	2251	33	3061	341
1533	511	2257	341	3067	1023
1541	511	2305	1023	3103	1023
1553	511	2311	341	3117	1023
1555	511	2327	1023	3121	341
1563	511	2347	1023	3133	1023
1577	511	2355	341	3171	1023
1605	511	2363	1023	3177	1023
1617	511	2377	1023	3205	93
1641	73	2413	93	3211	1023
1665	511	2415	1023	3247	93
1671	511	2431	1023	3255	341
1707	511	2437	341	3265	1023
1713	511	2443	1023	3277	341
1715	511	2461	1023	3301	1023
1725	511	2475	1023	3315	341
1731	511	2503	1023	3323	1023
1743	511	2527	1023	3337	1023
1751	511	2541	93	3367	341

Table III-5   (Cont'd).

Degree (bold) & polynomial	Period	Degree (bold) & polynomial	Period	Degree (bold) & polynomial	Period
1023	3375	4161	2047	4767	2047
341	3417	4173	2047	5001	2047
341	3421	4215	2047	5007	2047
1023	3427	4225	2047	5023	2047
1023	3435	4237	2047	5025	2047
1023	3441	4251	2047	5051	2047
3453	93	4261	2047	5111	2047
3465	341	4303	89	5141	2047
3471	1023	4317	2047	5155	2047
3507	1023	4321	2047	5171	2047
3515	1023	4341	2047	5177	2047
3525	1023	4347	2047	5205	2047
3531	1023	4353	2047	5221	2047
3543	1023	4365	2047	5235	2047
3573	341	4415	2047	5247	2047
3575	1023	4423	2047	5253	2047
3601	341	4445	2047	5263	2047
3607	341	4451	2047	5265	2047
3615	1023	4467	89	5325	2047
3623	1023	4473	2047	5337	2047
3651	341	4475	2047	5343	23
3661	1023	4505	2047	5351	2047
3705	341	4511	2047	5357	2047
3733	1023	4521	2047	5361	2047
3753	341	4533	2047	5373	2047
3763	1023	4563	2047	5403	2047
3771	1023	4565	2047	5411	2047
3777	11	4577	2047	5421	2047
		4603	2047	5463	2047
11		4617	2047	5477	2047
4005	2047	4653	2047	5501	2047
4027	2047	4655	2047	5513	2047
4053	2047	4671	2047	5531	2047
4055	2047	4707	2047	5537	2047
4107	2047	4731	2047	5545	2047
4143	2047	4745	2047	5557	2047
4145	2047	4757	89	5575	2047

Table III-5 (Cont'd).

Degree (bold) & polynomial	Period	Degree (bold) & polynomial	Period	Degree (bold) & polynomial	Period
5607	2047	6367	2047	7137	2047
5613	2047	6403	2047	7161	2047
5623	2047	6417	2047	7173	2047
5625	2047	6435	2047	7175	2047
5657	2047	6447	2047	7201	2047
5667	2047	6455	2047	7223	2047
5675	2047	6501	2047	7237	2047
5711	2047	6507	2047	7243	2047
5733	2047	6525	2047	7273	2047
5735	2047	6531	2047	7311	89
5747	2047	6543	2047	7317	2047
5755	2047	6557	2047	7335	2047
6013	2047	6561	2047	7363	2047
6015	2047	6623	2047	7371	2047
6031	2047	6637	2047	7413	2047
6037	2047	6651	2047	7431	2047
6061	89	6673	2047	7461	2047
6127	2047	6675	2047	7467	2047
6141	2047	6711	2047	7535	2047
6153	2047	6727	2047	7553	2047
6163	2047	6733	2047	7555	2047
6165	23	6741	2047	7565	2047
6205	2047	6747	2047	7571	89
6211	2047	6765	2047	7603	2047
6227	2047	6777	89	7621	2047
6233	2047	7005	2047	7627	2047
6235	2047	7035	2047	7633	2047
6263	2047	7041	2047	7647	2047
6277	2047	7047	2047	7655	2047
6307	2047	7053	2047	7665	2047
6315	2047	7063	2047	7715	2047
6323	2047	7071	2047	7723	2047
6325	2047	7107	2047	7745	2047
6343	2047	7113	2047	7751	2047
6351	2047	7125	2047	7773	89

5. Reduce the exponents of the remaining polynomials modulo 7. If the result appears in (2), the original polynomial is to be crossed off the list. This eliminates 90 polynomials, leaving 186 on the list.

Since $\Psi_2(11) = 186$, the list is now a table of the 11th degree polynomials irreducible mod 2. However, $\lambda_2(11) = 176$, and some auxiliary method must still be employed to determine which 10 do not correspond to maximum-length shift register sequences.

The computation of the irreducible polynomials modulo 2 was programmed for the higher degrees appearing in Table V at Lincoln Laboratories, for the Memory Test Computer (M. T. C.). The list was printed *octally*, in accordance with the following code:

$$x^{14} \quad x^{13} \quad x^{12} \quad x^{11} \quad x^{10} \quad x^9 \quad x^8 \quad x^7 \quad x^6 \quad x^5 \quad x^4 \quad x^3 \quad x^2 \quad x \quad 1$$
$$1 \quad 1 \quad 1 \quad 1 \quad 1 \quad 1 \quad 1 \quad 1 \quad 1 \quad 1 \quad 1 \quad 1 \quad 1 \quad 1 \quad 1. \tag{3}$$

For example, the polynomial $x^{12} + x^9 + x^7 + x^2 + 1$ becomes /001/000/ 110/000/101/, and taking the binary numbers in blocks of 3, this is 10605, which is the way the machine would print it.

Conversely, given the octal number 21375, this is the binary number /010/001/011/111/101/, representing the polynomial $x^{13} + x^9 + x^7 + x^6 + x^5 + x^4 + x^3 + x^2 + 1$.

6.5. *The synthetic approach*

As pointed out in Section 5, every maximum-length shift register sequence can be obtained as a superposition of cyclotomic cosets. Then, from the sequence, it is comparatively easy to work back to the polynomial.

For example, to find the six "maximum-period" polynomials of degree 5, it is first necessary to list the cosets:

$$
\begin{array}{llllll}
C_0: & 1, & 2, & 4, & 8, & 16 \\
C_1: & 3, & 6, & 12, & 24, & 17 \\
C_2: & 9, & 18, & 5, & 10, & 20 \\
C_3: & 27, & 23, & 15, & 30, & 29 \\
C_4: & 19, & 7, & 14, & 28, & 25 \\
C_5: & 26, & 21, & 11, & 22, & 13 \,.
\end{array}
\tag{4}
$$

These are obtained by using the powers of 2 for C_0; using any number which has not yet appeared, times the elements of C_0 (mod 31) as C_1; and similarly for the other cosets.

Next, it is necessary to assign 1's and 0's to the cosets. From Reuschle's Table [31], this can be done as in 5.12, to yield six maximum-length shift register sequences. From these, the recurrence relations can be ascertained, and hence the characteristic polynomials.

A more difficult example is the case of sequences of length 15. The cosets are

$$
\begin{aligned}
C_0&: \ 1, \ \ 2, \ \ 4, \ \ 8 \\
C_1&: \ 7, \ 14, \ 13, \ 11 \\
C_a&: \ 3, \ \ 6, \ 12, \ \ 9 \\
C_b&: \ 5, \ 10, \ \ 5, \ 10 \\
C_c&: \ 0, \ \ 0, \ \ 0, \ \ 0 .
\end{aligned}
\qquad (5)
$$

Here C_0 and C_1 are the "true cosets," in that their elements constitute the multiplicative group mod 15. It is for them that the values 0 and 1 must be substituted judiciously. The other seven numbers will be the same in all sequences. Since there must be one more 1 than 0, C_a must always have the value 1, while C_b and C_c always take the value 0. The "table" for C_0 and C_1 is

$$
\begin{array}{c c c}
 & C_0 & C_1 \\
S_1 & 1 & 0 \\
S_2 & 0 & 1 ,
\end{array}
\qquad (6)
$$

which is always the case when there are only two true cosets.

The sequences of length 15 are shown in Table III-6:

Table III-6.

	Fixed	C_0	C_1	S_1	S_2
0	0			0	0
1		✓		1	0
2		✓		1	0
3	1			1	1
4		✓		1	0
5	0			0	0
6	1			1	1
7			x	0	1
8				1	0
9	1			1	1
10	0			0	0
11			x	0	1
12	1			1	1
13			x	0	1
14			x	0	1

To find the polynomial corresponding to S_1, assume that S_1 satisfies the recurrence

$$a_n = c_1 a_{n-1} + c_2 a_{n-2} + c_3 a_{n-3} + c_4 a_{n-4}. \tag{7}$$

From the sequence itself, one obtains the relations

$$\begin{aligned}
1 &= c_2 + c_3 + c_4 \\
0 &= c_1 + c_2 + c_3 + c_4 \\
1 &= c_1 + c_2 + c_3 \\
0 &= c_1 + c_2 + c_4.
\end{aligned} \tag{8}$$

From the first two, $c_1 = 1$. From the second and fourth, $c_3 = 0$. Then, from the third, $c_2 = 0$, and finally, $c_4 = 1$. Thus S_1 satisfies

$$a_n = a_{n-1} + a_{n-4}, \tag{9}$$

and has

$$f(x) = 1 + x + x^4 \tag{10}$$

as its characteristic polynomial. Similarly, $f(x) = 1 + x^3 + x^4$ is the characteristic polynomial of S_2.

6.6. *Evaluating the cosets*

Reuschle's Table [31] is neither extensive nor accurate. Hence, it is well to include some remarks on the "evaluation" of the cyclotomic cosets, that is, determining how to assign 1's and 0's to them.

The basic device here is "Gauss's product formula" [39] which expresses the "product" of any two cosets as a sum of cosets. (This "product" is in terms of evaluation with 1's and 0's. It has nothing to do with multiplication in the factor group.) Specifically,

$$C_i C_j = \sum_{v \text{ in } \sigma_j} C^{(u+v)}, \tag{11}$$

where $C^{(m)}$ means the coset containing the number m, where u is a fixed element of C_i, and v is a variable element of C_j. (This product is independent of the choice of u in C_i; and $C_i C_j = C_j C_i$.)

For example, to compute $C_0 C_1$ from (4) by Gauss's formula, take $u = 1$ and $v = 3, 6, 12, 24, 17$. Then $u + v = 4, 7, 13, 25, 18$, belonging to the cosets C_0, C_4, C_3, C_4 and C_2, respectively. Working mod 2,

$$C_0 C_1 = C_0 + C_2 + C_3. \tag{12}$$

If ever $u + v = 0$, $C^{(0)} = r$, the degree of the polynomials being determined.

In this fashion

$$C_0C_1 = C_0 + C_2 + C_3$$
$$C_0C_2 = C_1 + C_3 + C_4$$
$$C_0C_3 = 5 + C_0 + C_1 + C_3 + C_4$$
$$C_0C_4 = C_0 + C_2 + C_3$$
$$C_0C_5 = C_1 + C_4 + C_5$$
$$C_0C_6 = C_0 .$$

(13)

For convenience, take $C_0 = 0$. Then (13) becomes

$$0 = C_2 + C_3$$
$$0 = C_1 + C_2 + C_4$$
$$0 = 1 + C_1 + C_3 + C_4$$
$$0 = C_1 + C_4 + C_5 .$$

(14)

Since the first equation is the sum of the second and fourth, there are only three equations in five unknown. Thus, it can do no harm to let, say, $C_0 = C_2 = C_5 = 0$. Then to preserve the balance of 1's and 0's, $C_1 = C_3 = C_4 = 1$.

This leads to the correct evluation:

	C_0	C_1	C_2	C_3	C_4	C_5
S_1	0	1	0	1	1	0
S_2	0	0	1	0	1	1
S_3	1	0	0	1	0	1
S_4	1	1	0	0	1	0
S_5	0	1	1	0	0	1
S_6	1	0	1	1	0	0 .

(15)

6.7. Supplementing the sieve method

The synthetic method can be used not only to determine directly the "maximum-length" polynomials of a given degree, but also to ascertain which irreducible polynomials of a given degree do *not* correspond to maximum period. That is, the synthetic method may be used as a *supplement* to the sieve method in the computation of polynomials of maximum exponent.

For example, there are 186 irreducible polynomials of degree 11, of which 176 have maximum exponent, 2047. Of the other 10, two have exponent 23 and eight have exponent 89 (see 3.6). To find the two with period 23, list the cosets mod 23:

$$C_0: \quad 1, \quad 2, \quad 4, \quad 8, \quad 16, \quad 9, \quad 18, \quad 13, \quad 3, \quad 6, \quad 12$$
$$C_1: \quad 5, \quad 10, \quad 20, \quad 17, \quad 11, \quad 22, \quad 21, \quad 19, \quad 15, \quad 7, \quad 14 .$$

(16)

From the table (6), the two shift register sequences of period 23 corresponding to irreducible 11th degree polynomials are

	C_0	C_1	S_1	S_2		C_0	C_1	S_1	S_2
0			1	1	12	✓		1	0
1	✓		1	0	13	✓		1	0
2	✓		1	0	14		x	0	1
3	✓		1	0	15		x	0	1
4	✓		1	0	16	✓		1	0
5		x	0	1	17		x	0	1
6	✓		1	0	18	✓		1	0
7		x	0	1	19		x	0	1
8	✓		1	0	20		x	0	1
9	✓		1	0	21		x	0	1
10		x	0	1	22		x	0	1
11		x	0	1					

$$(17)$$

To find the polynomials, assume a recursion relation for S_1:

$$a_n = c_1 a_{n-1} + c_2 a_{n-2} + \cdots + c_{11} a_{n-11} . \tag{18}$$

Directly from (17), one obtains

$$0 = c_2 + c_3 + c_5 + c_7 + c_8 + c_9 + c_{10} + c_{11}$$
$$1 = c_3 + c_4 + c_6 + c_8 + c_9 + c_{10} + c_{11}$$
$$1 = c_1 + c_4 + c_5 + c_7 + c_9 + c_{10} + c_{11}$$
$$0 = c_1 + c_2 + c_5 + c_6 + c_8 + c_{10} + c_{11}$$
$$0 = c_2 + c_3 + c_6 + c_7 + c_9 + c_{11}$$
$$1 = c_3 + c_4 + c_7 + c_8 + c_{10}$$
$$0 = c_1 + c_4 + c_5 + c_8 + c_9 + c_{11}$$
$$1 = c_2 + c_5 + c_6 + c_9 + c_{10}$$
$$0 = c_1 + c_3 + c_6 + c_7 + c_{10} + c_{11}$$
$$0 = c_2 + c_4 + c_7 + c_8 + c_{11}$$
$$0 = c_3 + c_5 + c_8 + c_9$$
$$0 = c_4 + c_6 + c_9 + c_{10} .$$

$$(19)$$

Solving simultaneously,

$$1 = c_1 = c_5 = c_6 = c_7 = c_9 = c_{11} ,$$

while

$$0 = c_2 = c_3 = c_4 = c_8 = c_{10} .$$

Thus the recursion formula is

$$a_n = a_{n-1} + a_{n-3} + a_{n-6} + a_{n-7} + a_{n-9} + a_{n-11} , \tag{20}$$

and the corresponding polynomial is

$$f_1(x) = x^{11} + x^9 + x^7 + x^6 + x^5 + x + 1 . \tag{21}$$

In like fashion, the polynomial for S_2 is

$$f_2(x) = x^{11} + x^{10} + x^6 + x^5 + x^4 + x^2 + 1 . \tag{22}$$

It is only slightly more difficult to determine the eight polynomials with period 89. This has been done in Table III-5.

6.8. *Direct factorization*

The *synthetic method* as it has just been presented is far simpler than the more "direct" form which will be outlined in this section. The main purpose of this section is to show that without the principle of the superposition of cosets it is still possible to execute the factorization of cyclotomic polynomials, but that even in the most favorable cases, more work is involved than in the "superposition" form of the method.

A program for the "direct method" is as follows:

To find the irreducible polynomials of degree r and exponent p (where it is *not* required that $p = 2^r - 1$),

1. List the *cyclically distinct* binary numbers with r binary digits, grouped by the number of 1's occurring. "All 0's" and "all 1's" may be omitted. (1 0 0 1 and 0 0 1 1 are *not* cyclically distinct; 1 0 0 1 and 0 1 0 1 are.)

2. Write the decimal equivalents of the binary numbers, mod p.

3. Find the table of cosets, and the table of evaluations, corresponding to the period p.

4. Replace each decimal equivalent by the evaluation of the coset to which it belongs.

5. Sum the evaluations for each grouping.
(Grouping is by the number of 1's occurring.) Then $x^r + \sigma_1 x^{r-1} + \sigma_2 x^{r-2} + \cdots + 1$ is a polynomial of the desired type, where σ_1 is the sum from the first grouping (one 1), σ_2 the sum from the second grouping (two 1's), etc.

In step 4, there are $\phi(p)/r$ choices of evaluations, and these yield *all* the polynomials of the type indicated.

As an example, to determine the irreducible polynomials of period 31 and degree 5.

1. The cyclically distinct binary numbers are

$$
\begin{array}{ccccc}
0 & 0 & 0 & 0 & 1 \\
\hline
0 & 0 & 0 & 1 & 1 \\
0 & 0 & 1 & 0 & 1 \\
\hline
0 & 0 & 1 & 1 & 1 \\
0 & 1 & 0 & 1 & 1 \\
\hline
0 & 1 & 1 & 1 & 1.
\end{array}
\tag{23}
$$

2. The decimal values are

$$
\begin{array}{c}
1 \\
\hline
3,\ 5 \\
\hline
7,\ 11 \\
\hline
15\ .
\end{array}
\tag{24}
$$

3. From (4), the corresponding cosets are

$$
\begin{array}{l}
\sigma_1 \quad C_0 \\
\sigma_2 \quad C_1,\ C_2 \\
\sigma_3 \quad C_4,\ C_3 \\
\sigma_4 \quad C_3\ .
\end{array}
\tag{25}
$$

4. From (15), the six evaluations are

$$
\begin{array}{ccccccc}
 & f_1 & f_2 & f_3 & f_4 & f_5 & f_6 \\
\sigma_1 & 0 & 0 & 1 & 1 & 0 & 1 \\
\sigma_2 & 1 & 1 & 0 & 1 & 0 & 1 \\
\sigma_3 & 1 & 0 & 1 & 1 & 1 & 0 \\
\sigma_4 & 1 & 0 & 1 & 0 & 0 & 1\ .
\end{array}
\tag{26}
$$

5. Writing $f(x) = x^5 + \sigma_1 x^4 + \sigma_2 x^3 + \sigma_3 x^2 + \sigma_4 x + 1$, the six polynomials in question are

$$
\begin{aligned}
f_1(x) &= x^5 + x^3 + x^2 + x + 1 \\
f_2(x) &= x^5 + x^3 + 1 \\
f_3(x) &= x^5 + x^4 + x^2 + x + 1 \\
f_4(x) &= x^5 + x^4 + x^3 + x^2 + 1 \\
f_5(x) &= x^5 + x^3 + 1 \\
f_6(x) &= x^5 + x^4 + x^3 + x + 1\ .
\end{aligned}
\tag{27}
$$

The table of cosets, and the corresponding evaluations, are required both for the superposition method and the direct method. However, the direct method involves listing the cyclically distinct binary numbers with r bits, which already requires several pages in the case $p = 23$,

$r = 11$ (done more quickly in 6.7 by the superposition method). In the direct method, it is also necessary to weight binary numbers less heavily if they have symmetry: 010101 has 3-fold symmetry, and must be weighted by $1/3$; 011011 has 2-fold symmetry, and must be weighted by $1/2$; etc.

The r simultaneous equations which must be solved at the end of the superposition method involve much less work than the evaluation of the binary digits (approximately $2^r/r$ of them), especially when r is large.

It is interesting that both forms of the synthetic method depend more critically on the size of r than on the size of p.

6.9. Useful tricks

Given an irreducible polynomial, it is often desired to find another, either of the same degree or of a higher degree. Several methods which accomplish this will be described here.

1. Given the irreducible polynomial $f(x) = \sum_{n=0}^{r} a_n x^n$, let $g(x) = \sum_{n=0}^{r} a_n x^{r-n}$. Then $g(x)$ is also irreducible. The shift register sequence corresponding to $g(x)$ is simply the time inverse of the sequence corresponding to $f(x)$. Thus, if $f(x)$ has maximum exponent, so does $g(x)$.

2. If $f(x)$ is irreducible, so is $h(x) = f(x + 1)$, since any factorization of $h(x)$ would factor $f(x)$. However, $h(x)$ may fail to have maximum exponent, even though $f(x)$ has. For example, if $f(x) = x^4 + x^3 + 1$ with $p = 15$, then $h(x) = (x + 1)^4 + (x + 1)^3 + 1 = x^4 + x^3 + x^2 + x + 1$, with $p = 5$.

3. Given a single irreducible polynomial of degree r, it is possible to get as many as five others by alternating methods 1 and 2. For example, from any one of the irreducible polynomials of degree 5, it is thus possible to obtain all six.

When $p = 2^r - 1$ is prime, all the irreducible polynomials necessarily have maximum exponent; and starting with a single maximum exponent polynomial, five others can always be obtained, if $r \geq 5$.

4. Suppose $f(x) = \sum_{n=0}^{r} a_n x^n$ is an irreducible polynomial (mod 2) of degree r, with maximum exponent. It can be shown that

$$F(x) = \sum a_n x^{2^r - 1} \tag{28}$$

is an *irreducible* polynomial of degree $2^r - 1$. However, it does not necessarily follow that $F(x)$ has maximum exponent. Of course, if $2^q - 1$ is *prime*, where $q = 2^r - 1$, then $F(x)$ *must* have maximum exponent.

For example, starting with $f(x) = x^2 + x + 1$, which is irreducible and has maximum exponent, one obtains $F(x) = x^3 + x + 1$, irreducible with maximum exponent (since 7 is prime). Applying the same transformation to $F(x)$ yields $G(x) = x^7 + x + 1$, irreducible with maximum exponent. Applying the transformation again, $H(x) = x^{127} + x + 1$ is irreducible with maximum exponent.

5. In 5.9 it was shown that "decimation" does not affect the property $R - 3$. The same reasoning can be used to prove the following:

> *Theorem 6.1.* If $\{a_n\}$ is a maximum-length shift register sequence with period p, and if $q_0, q_1, \ldots, q_l$ is any set of representatives of the cyclotomic cosets $C_0, C_1, \ldots, C_l$, then $\{a_{nq_0}\}, \{a_{nq_1}\}, \ldots, \{a_{nq_l}\}$ are *all* the maximum-length shift register sequences of period p.

Hence, given one maximum exponent polynomial of degree r, all the others can be gotten by writing out the corresponding sequence, decimating, and finding the polynomials by simultaneous equations.

Of course, only the *proper* cosets are referred to in Theorem 6.1., since q must be relatively prime to p if $\{a_{qn}\}$ is to have the same period as $\{a_n\}$.

6.10. *Summary*

There are two basic approaches to the computation of the polynomials corresponding to pseudo-noise sequences. One involves the *sieve method*, which eliminates the sequences which do not have maximum length. The other method is an application of the *superposition of cosets*, developed in Section 5.

In addition, several techniques are presented for finding new maximum exponent polynomials from given ones.

Chapter IV

STRUCTURAL PROPERTIES OF PN SEQUENCES

1. INTRODUCTION

For the purposes of this discussion, a PN sequence is defined to be a maximum-length linear recurring sequence modulo 2. That is, $\{a_k\}$ is a PN sequence if and only if it is a binary sequence which satisfies a linear recurrence

$$a_k = \sum_{i=1}^{n} c_i a_{k-i} \ (\text{modulo } 2) \tag{1}$$

and has period $p = 2^n - 1$. The number n is referred to as the degree of the PN sequence $\{a_k\}$.

The polynomial

$$f(x) = 1 + \sum c_i x^i \ (\text{modulo } 2) \tag{2}$$

is called the characteristic polynomial of the sequence $\{a_k\}$ of Equation (1). The irreducibility of $f(x)$ is a necessary condition for $\{a_k\}$ to be a PN sequence. The necessary and sufficient condition is that $f(x)$ divide $1 - x^m$ for $m = p$, but for no positive m less than p.

The number of distinct PN sequences of length $p = 2^n - 1$ (where two sequences differing only in their starting point are not considered distinct) is $\phi(p)/n$, where ϕ is Euler's function. Morever, the PN sequences have been characterized in Chapter III as precisely those binary sequences which possess the "cycle-and-add" property.[1]

Beyond these basic facts about PN sequences, there are certain underlying structural properties common to all PN sequences of given length $p = 2^n - 1$. These regularities are based on the cyclotomic cosets modulo p, which are defined in Section 3. This concept allows the construction of PN sequences by the method of *superposition of cosets*, as well as the formation of new PN sequences from a given sequence by the process of *decimation* (Sections 4 and 5). Also, the cross-correlation properties and Fourier series coefficients (Sections 6 and 7) are related to the coset decomposition. Finally, in Section 8,

---

[1] This is the property characterized by Theorems 4.3. and 4.5. on pages 44-47.

it is shown that all perfect sequences (sequences with two-level auto-correlation) can be studied by the coset method.

2. MULTIPLIERS OF PN SEQUENCES

Let D be the unit delay operator, so that $Da_k = a_{k-1}$, and $D^2 a_k = a_{k-2}$. Then by Equations (1) and (2),

$$f(D)\{a_k\} = 0 \qquad (3)$$

is equivalent to the recursion relation (Eq. 1) satisfied by $\{a_k\}$.

Among polynomials modulo 2, $f(x^2) = [f(x)]^2$, and more generally $f(x^{2^i}) = [f(x)]^{2^i}$, in view of the simplified binomial theorem

$$(a + b)^{2^i} = a^{2^i} + b^{2^i} \text{ (modulo 2)} . \qquad (4)$$

The values of j other than power of 2 do not, in general, satisfy $f(x^j) = [f(x)]^j$.

Consider the sequence $\{a_{2k}\}$, which is formed from $\{a_k\}$ by taking alternate terms. Thus

$$f(D^2)\{a_k\} = f(D)\{a_{2k}\} , \qquad (5)$$

where the two sequences differ by at most a fixed translation (phase shift). However, $f(D^2)\{a_k\} = f(D)\,[f(D)\{a_k\}] = f(D)\,\{0\} = \{0\}$. Hence

$$f(D)\{a_{2k}\} = 0 \qquad (6)$$

so that $\{a_{2k}\}$ satisfies the same recursion formula as $\{a_k\}$. Therefore, $\{a_{2k}\}$ is identical to $\{a_k\}$, except for a possible phase shift. Thus we have proved the following theorem.

Theorem 1. If $\{a_k\}$ is a PN sequence, then $\{a_{qk}\}$ equals $\{a_k\}$ except for a phase shift, when $q = 1, 2, 4, 8, \ldots, 2^{n-1}$.

The numbers $1, 2, 4, \ldots, 2^{n-1}$ are called the *multipliers* of the sequence $\{a_k\}$. Collectively, they are known as the multiplier group since they form a group under multiplication.

Example 1. Let $a_0 = 1$, $a_1 = 1$, $a_2 = 1$, $a_3 = 0$, $a_4 = 1$, $a_5 = 0$, $a_6 = 0$, which is a PN sequence of degree $n = 3$ and period $p = 2^3 - 1 = 7$. Thus,

$$a_k = 1110100$$
$$a_{2k} = 1110100$$
$$a_{4k} = 1110100 ,$$

while

$$a_{3k} = 1001011$$
$$a_{5k} = 1001011$$

$$a_{6k} = 1001011.$$

Example 2. Let $a_0 = 1$, $a_1 = 1$, $a_2 = 0$, $a_3 = 1$, $a_4 = 0$, $a_5 = 0$, $a_6 = 1$. Then

$$a_k = 1101001$$
$$a_{2k} = 1001110$$
$$a_{4k} = 1010011$$

are merely phase shifts of each other.

3. THE CYCLOTOMIC COSETS

There are $\phi(p)$ numbers from 1 to p which are relatively prime to p. (This is in fact the defining property of Euler's function ϕ.) The $\phi(p)$ numbers form a group under multiplication modulo p, and if p is odd, the set $(1, 2, 4, 8, \ldots)$ forms a subgroup. Specifically, with $p = 2^n - 1$, the multiplier subgroup consists of the n elements $(1, 2, 4, \ldots, 2^{n-1})$.

A coset is obtained by taking any element of the large group and multiplying it by each number of the subgroup in turn. For example, with $p = 31$, the cosets of the multiplier subgroup are

$$
\begin{array}{lccccc}
C_1: & 1 & 2 & 4 & 8 & 16 \\
C_2: & 3 & 6 & 12 & 24 & 17 \\
C_3: & 9 & 18 & 5 & 10 & 20 \\
C_4: & 27 & 23 & 15 & 30 & 29 \\
C_5: & 19 & 7 & 14 & 28 & 25 \\
C_6: & 26 & 21 & 11 & 22 & 13 \\
\end{array}
$$

Clearly the number of cosets is always $\phi(2^{n-1})/n$, which in the case of $n = 5$ yields $\phi(31)/5 = 6$.

In addition to these $\phi(p)/n$ *proper* cosets, there are always one or more *improper* cosets, or *generalized* cosets. These result from multiplying an integer which is not relatively prime to p times each element of the subgroup. Thus (0) is always a coset unto itself. In the modulo 31 example, (0) is the only improper coset. However, the coset decomposition modulo 15 is

$$
\begin{array}{lcccll}
C_1: & 1, & 2, & 4, & 8 & \text{(proper)} \\
C_2: & 7, & 14, & 13, & 11 & \text{(proper)} \\
C_3: & 3, & 6, & 12, & 9 & \text{(improper)} \\
C_4: & 5, & 10 & & & \text{(improper)} \\
C_5: & 0 & & & & \text{(improper)} .
\end{array}
$$

The set of all cosets (proper and improper) of the multiplier subgroup constitutes the cyclotomic cosets modulo p. The number of cyclotomic cosets modulo p, denoted by $Y(p)$, can be expressed in several ways:

$$Y(p) = \sum_{d \mid p} \frac{\phi(d)}{e_2(d)} = \frac{1}{n} \sum_{i=1}^{n} [2^{(i,n)} - 1] = \left[\frac{1}{n} \sum_{d \mid n} \phi(d) 2^{n/d} \right] - 1. \quad (7)$$

In particular, $Y(p) \geq (p + n - 1)/n$ for all n, and $Y(p)$ is odd for $n \neq 2$. (Cf. the discussion of $Z(n) = Y(p) + 1$ in Ch. VI, Section 2.)

4. DECIMATION OF SEQUENCES

It was shown in Section 2 that alternation of a PN sequence (that is, replacing $\{a_k\}$ by $\{a_{2k}\}$) does not alter the order of the terms, except perhaps the location of the starting point. More generally, $\{a_{qk}\}$ is the same as $\{a_k\}$ for $q = 1, 2, 4, \ldots, 2^{n-1}$. In this section, the behavior of $\{a_{qk}\}$ will be investigated with no restrictions on q. (The replacement of $\{a_k\}$ by $\{a_{qk}\}$ is termed *decimation*.)

> *Theorem 2.* If $\{a_k\}$ is a PN sequence with period p, then $\{a_{qk}\}$ is again a PN sequence, with the same period, if and only if $(q, p) = 1$. If both $(q_1, p) = 1$ and $(q_2, p) = 1$, then $\{a_{q_1 k}\} = \{a_{q_2 k}\}$ (except for the starting point) if and only if q_1 and q_2 belong to the same (proper) cyclotomic coset modulo p.

Proof. If $(q, p) > 1$, then $\{a_{qk}\}$ has period $p/(p, q) < p$.

If $(q, p) = 1$, then $\{a_{qk}\}$ is again a PN sequence because (1) its period remains p, and (2) from $f(D)\{a_k\} = 0$ it follows that $f(D^{1/q})\{a_{qk}\} = 0$, where $1/q$ is the multiplicative inverse of q modulo p. However, $f_1(D) = \prod_{i=1}^{n} f(\omega^i D^{1/q})$ is a polynomial of degree n, with its coefficients in the two-element field $(0, 1)$, which is divisible by $f(D^{1/q})$. (Here ω denotes a primitive q^{th} root of unity.) Hence, $f_1(D)\{a_{qk}\} = 0$, which shows that $\{a_{qk}\}$ satisfies a linear recurrence relation of degree n.

It remains only to show that $\{a_{q_1 k}\} \neq \{a_{q_2 k}\}$ when q_1 and q_2 are in different cyclotomic cosets. In view of Theorem 1, it suffices to show that $\{a_{qk}\} = \{a_k\}$ only if $q = 1, 2, 4, \ldots, 2^{n-1}$; that is, that only the powers of 2 are multipliers.

Consider the Galois field $GF(2^n)$ obtained by adjoining the roots of $f(x) = 0$ to $GF(2)$. If q is a multiplier, then $\omega \to \omega^q$ is an automorphism of $GF(2^n)$ over $GF(2)$. However, the number of automorphisms of an extension of degree n cannot exceed n, so that $\omega \to \omega^0$, $\omega \to \omega^1$, $\omega \to \omega^2$, $\ldots, \omega^{2^{n-1}}$ are the only automorphisms, and $1, 2, \ldots, 2^{n-1}$ are the only multipliers.

Corollary 1. All $\phi(2^n - 1)/n$ PN sequences of length $p = 2^n - 1$ can be obtained from any given one by suitable decimations.

Corollary 2. If q is a primitive element modulo p, then all $\phi(2^n - 1) \cdot n$ characteristic polynomials for the PN sequences of length p can be obtained from a given one $f_1(x)$ by

$$f_{k+1}(x) = \prod_{i=1}^{q} f_k(\omega^i x^{1/q}), \quad 1 \leq k \leq \phi(2^n - 1) \cdot n, \tag{8}$$

where ω is a primitive root of $x^q = 1$.

Corollary 3. Taking $q = 3$ in Equation (8) leads to *tertiation* of the PN sequence. This is a new PN sequence if and only if n is odd. The number of distinct PN sequences obtainable by tertiation is the smallest j such that $3^j \equiv 2^m$ (modulo p) for any m.

It may be noted that if $f(x)$ is the characteristic polynominal of a PN sequence, then $f(x)$ will divide $f(x^t)$ only if $t = 1, 2, 4, 8, \ldots$. For other polynomials $f(x)$, however, $f(x)$ may divide $f(x^t)$ for a wider range of values. This is true even for irreducible polynomials, such as $f(x) = x^4 + x^3 + x^2 + x + 1$, which divides $f(x^3)$. This implies that the sequence 11110, repeating with period 5, which has $f(x) = x^4 + x^3 + x^2 + x + 1$ as its characteristic polynomial, is "decimated into itself" by $q = 3$ as well as by $q = 1, 2, 4$.

5. THE SUPERPOSITION OF COSETS

Let $\{a_k\}$ be a PN sequence. Then $\{a_{2k}\}$ is simply a phase shift of $\{a_k\}$ (by Theorem 1) so that termwise

$$\{a_{2k}\} = \{a_{k+\tau}\}$$

for some τ. In this section, the notation $\{a_k\} = \{b_k\}$ will be restricted to denote term-by-term equality.

Lemma. There is a phase shift $\{b_k\} = \{a_{k+m}\}$ of $\{a_k\}$, for some m, such that $\{b_{2k}\} = \{b_k\}$.

Proof. Suppose originally that $\{a_{2k}\} = \{a_{k-\tau}\}$. Choose $m = 2\tau$, so that $k' = k + 2\tau$. Then

$$\{b_{2k}\} = \{a_{2k+m}\} = \{a_{2k+2\tau}\} = \{a_{2(k+\tau)}\} = \{a_{(k+\tau)-\tau}\} = \{a_{k-m}\} = \{b_k\}.$$

Example. If $a_0 = 0$, $a_1 = 0$, $a_2 = 1$, $a_3 = 1$, $a_4 = 1$, $a_5 = 0$, $a_6 = 1$, then $\{a_{2k}\} = \{0111010\} = \{a_{k+1}\}$. Hence $\{b_k\} = \{a_{k-2}\} = \{1110100\}$ satisfies $\{b_{2k}\} = \{b_k\}$.

Theorem 3. Every PN sequence $\{a_k\}$ has a phase shift $\{b_k\}$ such that the value of b_k depends only on the cyclotomic coset to which k belongs modulo p, and not on the exact value of k.

Proof. Select $\{b_k\}$ as in the lemma, so that $\{b_{2k}\} = \{b_k\}$. In view of Theorem 2, $\{b_{q_1 k}\} = \{b_{q_2 k}\}$ (as a term-by-term equality) whenever q_1 and q_2 belong to the same cyclotomic coset. Specifically, in $\{b_k\}$,

$$b_1 = b_2 = b_4 = b_8 = \cdots$$
$$b_3 = b_6 = b_{12} = b_{24} = \cdots$$
$$b_9 = b_{29} = b_{49} = b_{89} = \cdots,$$

and the value of $\{b_k\}$ depends only on the cyclotomic coset to which k belongs.

Example 1. The sequence 1110100, with $p = 7$, is left termwise invariant by the multipliers $C_1 = (1, 2, 4)$, but is inverted by $C_2 = (3, 6, 5)$. It is seen that

$$b_0 = 1$$
$$b_1 = b_2 = b_4 = 1$$
$$b_3 = b_6 = b_5 = 0.$$

Example 2. To find all six PN sequences of degree $n = 5$ and length $p = 31$, we first list all the cyclotomic cosets:

$$
\begin{array}{lrrrrr}
C_0: & 0 \\
C_1: & 1, & 2, & 4, & 8, & 16 \\
C_2: & 3, & 6, & 12, & 24, & 17 \\
C_3: & 9, & 18, & 5, & 10, & 20 \\
C_4: & 27, & 23, & 15, & 30, & 29 \\
C_5: & 19, & 7, & 14, & 28, & 25 \\
C_6: & 26, & 21, & 11, & 22, & 13.
\end{array}
$$

The next problem is how to assign the values 0 and 1 to the various cosets. This can be determined by a method invented by Kummer [22] and tabulated by Reuschle [31]. The six assignments valid for the present case are

	C_0	C_1	C_2	C_3	C_4	C_5	C_6
S_1	1	0	1	1	0	0	1
S_2	1	1	0	1	1	0	0
S_3	1	0	1	0	1	1	0
S_4	1	0	0	1	0	1	1
S_5	1	1	0	0	1	0	1
S_6	1	1	1	0	0	1	0.

The six PN sequences which result are shown in Table IV·1.

Table IV·1. The six PN sequences of length 31 as superpositions of cosets.

k	C_0	C_1	C_2	C_3	C_4	C_5	C_6	S_1	S_2	S_3	S_4	S_5	S_6
0	1							1	1	1	1	1	1
1		1						0	1	0	0	1	1
2		1						0	1	0	0	1	1
3			1					1	0	1	0	0	1
4		1						0	1	0	0	1	1
5				1				1	1	0	1	0	0
6			1					1	0	1	0	0	1
7						1		0	0	1	1	0	1
8		1						0	1	0	0	1	1
9				1				1	1	0	1	0	0
10				1				1	1	0	1	0	0
11							1	1	0	0	1	1	0
12			1					1	0	1	0	0	1
13							1	1	0	0	1	1	0
14						1		0	0	1	1	0	1
15					1			0	1	1	0	1	0
16		1						0	1	0	0	1	1
17			1					1	0	1	0	0	1
18				1				1	1	0	1	0	0
19						1		0	0	1	1	0	1
20				1				1	1	0	1	0	0
21							1	1	0	0	1	1	0
22							1	1	0	0	1	1	0
23					1			0	1	1	0	1	0
24			1					1	0	1	0	0	1
25						1		0	0	1	1	0	1
26							1	1	0	0	1	1	0
27					1			0	1	1	0	1	0
28						1		0	0	1	1	0	1
29					1			0	1	1	0	1	0
30					1			0	1	1	0	1	0

The cosets thus appear as master switches, each of which regulates an entire bank of sequence positions on an all-or-none basis. This is the *principle of the superposition of cosets*.

Every $\{a_k\}$ has been shown to have a phase shift $\{b_k\}$ which satisfies $\{b_{2k}\} = \{b_k\}$. The term b_0 may be called the absolute zero term of $\{a_k\}$, since it remains invariant under decimation. Moreover, b_0 is unique, because it is always the term corresponding to the one-element cyclotomic coset (0).

6. REGULARITIES IN THE CROSS-CORRELATION OF PN SEQUENCES

By the results of the preceding section, all PN sequences of given length $p = 2^n - 1$ are superpositions of the same $Y(n)$ cyclotomic cosets modulo p. Hence the cross-correlation between any two such sequences will exhibit certain regularities. Given two sequences $\{a_k\}$ and $\{b_k\}$, both of period p, their cross-correlation function $C(\tau)$ is defined to be

$$C(\tau) = \sum_{k=1}^{p} a_k b_{k+\tau} . \tag{9}$$

In the special case that $\{a_k\} = \{b_k\}$, $C(\tau)$ is called the auto-correlation of the sequence $\{a_k\}$. For every PN sequence, the auto-correlation satisfies

$$C(\tau) = \begin{cases} \dfrac{p+1}{2} & \text{if } \tau = 0 \\[2mm] \dfrac{p+1}{4} & \text{if } 0 < \tau < p . \end{cases} \tag{10}$$

That is, PN sequences always possess a two-level auto-correlation. (This is a direct consequence of the delay-and-add property of PN sequences. Cf. Theorem 6 in Section 8.) For the case of cross-correlation, the only statement which holds in complete generality is the following theorem.

Theorem 4. The number of distinct values assumed by the cross-correlation function $C(\tau)$ of two PN sequences $\{a_k\}$ and $\{b_k\}$, each of period p, can never exceed $Y(p)$, the number of cyclotomic cosets modulo p.

Proof. Without loss of generality, arrange $\{a_k\}$ and $\{b_k\}$ to satisfy $\{a_{2k}\} = \{a_k\}$ and $\{b_{2k}\} = \{b_k\}$ as termwise equalities. Moreover, by Theorem 2, $\{b_k\} = \{a_{qk}\}$ for some q, with $(q, p) = 1$. Then

$$C(\tau) = \sum_{k=1}^{p} a_k b_{k+\tau} = \sum_{k=1}^{p} a_k a_{qk+\tau} . \tag{11}$$

By the formula of Gauss [39, Vol. I, p. 166] for the multiplication of cyclotomic cosets (or *periods*, as Gauss termed them), the value of $C(\tau)$ in Equation (11) depends only on the coset to which τ belongs, and not on its precise numerical value. Hence the number of different values assumed by $C(\tau)$ does not exceed the number of distinct cosets $Y(p)$.

Experimentally, far stronger regularities than any implied by Theorem 4 have been observed. In the case that p is prime, it is a common occurrence that $C(\tau)$ assumes only three distinct values. The full $Y(p)$ distinct values seem never to be attained or even closely approached unless the two sequences in question, $\{a_k\}$ and $\{b_k\}$, are time-inverses of each other; that is, unless $\{a_k\} = \{b_{p-k}\}$.

Example 1. $p = 7$. The two PN sequences are 1110100 and 1001011. Here

$$C(0) = 1$$
$$C(1) = C(2) = C(4) = 2$$
$$C(3) = C(6) = C(5) = 3 .$$

Example 2. $p = 15$. The two PN sequences are 011110101100100 and 000100110101111. This time

$$C(0) = 4$$
$$C(1) = C(2) = C(4) = C(8) = 3$$
$$C(7) = C(14) = C(13) = C(11) = 4$$
$$C(3) = C(6) = C(12) = C(9) = 5$$
$$C(5) = C(10) = 6 .$$

Example 3. $p = 31$. The six PN sequences are shown in Table IV-1. The only essentially distinct cases of cross-correlation involve S_1 vs S_2, S_1 vs S_3, and S_1 vs S_4.

(a) S_1 vs S_2

$$C(\tau) = \begin{cases} 6 \text{ for } \tau \text{ in } C_0, \ C_1 \\ 8 \text{ for } \tau \text{ in } C_1, \ C_5, \ C_6 \\ 10 \text{ for } \tau \text{ in } C_3, \ C_4 \end{cases}$$

(b) S_1 vs S_3

$$C(\tau) = \begin{cases} 6 \text{ for } \tau \text{ in } C_0, \ C_4 \\ 8 \text{ for } \tau \text{ in } C_1, \ C_3, \ C_6 \\ 10 \text{ for } \tau \text{ in } C_2, \ C_5 \end{cases}$$

(c) S_1 vs S_4

$$C(\tau) = \begin{cases} 11 \text{ for } \tau \text{ in } C_0 \\ 10 \text{ for } \tau \text{ in } C_1 \\ 9 \text{ for } \tau \text{ in } C_3, \ C_5 \\ 8 \text{ for } \tau \text{ in } C_4 \\ 7 \text{ for } \tau \text{ in } C_6 \\ 6 \text{ for } \tau \text{ in } C_2 . \end{cases}$$

Example 4. $p = 127$. There are 18 PN sequences of length 127. In each essentially different case, the cross-correlation was computed. The number of distinct values assumed by $C(\tau)$ was the following:

- (a) S_1 vs S_1 — 2 values of $C(\tau)$
- (b) S_1 vs S_2 — 3 values of $C(\tau)$
- (c) S_1 vs S_3 — 3 values of $C(\tau)$
- (d) S_1 vs S_4 — 3 values of $C(\tau)$
- (e) S_1 vs S_5 — 3 values of $C(\tau)$
- (f) S_1 vs S_6 — 3 values of $C(\tau)$
- (g) S_1 vs S_7 — 7 values of $C(\tau)$
- (h) S_1 vs S_8 — 7 values of $C(\tau)$
- (i) S_1 vs S_9 — 7 values of $C(\tau)$
- (j) S_1 vs S_{10} — 11 values of $C(\tau)$.

In case (j), consecutive integers occurred as values of $C(\tau)$. In

Table IV-2. The number of cyclotomic cosets
as a function of the period.

n	p	$Y(p)$
1	1	1
2	3	2
3	7	3
4	15	5
5	31	7
6	63	13
7	127	19
8	255	35
9	511	59
10	1023	107
11	2047	187
12	4095	351
13	8191	631
14	16,383	1181
15	32,767	2191
16	65,535	4115
17	131.071	7711
18	262,143	14,601
19	524,287	27,595
20	1,048,575	52,487

cases (g), (h), and (i), only even integers occurred as values of $C(\tau)$. In cases (b), (c), (d) (e), and (f), the values were all multiples of four; whereas in case (a) they were multiples of eight. By Theorem 4, the number of values of $C(\tau)$ for $p = 127$ could not exceed $Y(127) = 19$. (For $Y(p)$, see Table IV-2.)

A check on the accuracy of the correlation computation is provided by the formula

$$\sum_{r=1}^{p} C(r) = \left(\frac{p+1}{2}\right)^2 . \tag{12}$$

It may be possible to discover further regularities in the cross-correlation based on the work of Kummer [22].

7. THE FOURIER ANALYSIS OF PN SEQUENCES

Let $\{a_j\}$ be a PN sequence of period p. Physically, we may regard $\{a_j\}$ as a function of the continuous variable j which vanishes for all non-integral values of j as well as those integers j for which $a_j = 0$, but behaves as a unit impulse (or delta function) for integer values of j such that $a_j = 1$.

Based on its periodicity, $\{a_j\}$ admits of a Fourier series expansion. The Fourier coefficients are given by

$$c_k = \frac{1}{p} \sum_{j=1}^{p} a_j e^{2\pi i k j/p} . \tag{13}$$

It is easy to relate the magnitudes of the c_k to the auto-correlation $C(\tau)$ of $\{a_j\}$. Specifically,

$$|c_k|^2 = \frac{1}{p^2} \sum_{j=1}^{p} a_j e^{2\pi i k j/p} \sum_{m=1}^{p} a_m e^{-2\pi i k m/p} = \frac{1}{p^2} \sum_{j=1}^{p} \sum_{m=1}^{p} a_j a_m e^{2\pi i k(j-m)/p}$$

$$= \frac{1}{p^2} \sum_{m=1}^{p} \left(\sum_{r=1}^{p} a_m a_{m+r} \right) e^{2\pi i k r/p} = \frac{1}{p^2} \sum_{r=1}^{p} C(\tau) e^{2\pi i k \tau/p}$$

$$= \begin{cases} \frac{1}{p^2}[C(0) + (p-1)C(1)] & \text{if } p|k \\ \frac{1}{p^2}[C(0) - C(1))] & \text{if } p \nmid k . \end{cases}$$

This expression is based on the fact that $\{a_j\}$ has one correlation value $C(0)$ in phase, and another value $C(1)$ out of phase. Since no other assumptions have been made about $\{a_j\}$, the basic results of this section hold for the sequences of Section 8 also.

Using Equation (10) to describe $C(\tau)$,

$$|c_k| = \begin{cases} \dfrac{p+1}{2p} & \text{if } p|k \\[2mm] \dfrac{\sqrt{p+1}}{2p} & \text{if } p\nmid k . \end{cases} \tag{14}$$

If $\{a_j\}$ is interpreted as a train of square waves rather than as a train of impulses, it is necessary to multiply the Fourier coefficients in Equation (14) by the square-wave transform function

$$\frac{\sin(\pi k/p)}{\pi k/p} \ .$$

If $\{a_j\}$ were a PN sequence whose two states are represented by 1 and -1 (instead of 1 and 0), then Equation (14) would become

$$|c_k'| = \begin{cases} \dfrac{1}{p} & \text{if } p|k \\[2mm] \dfrac{\sqrt{p+1}}{p} & \text{if } p\nmid k . \end{cases} \tag{15}$$

The c_k' could also be multiplied by $\sin(\pi k/p)/(\pi k/p)$ to convert from impulses to square waves.

In addition to these results on the magnitudes of the Fourier coefficients, it is also possible to say something about the phases.

    Theorem 5. Let $\{a_j\}$ be phased so that $\{a_{2j}\} = \{a_j\}$. Then $c_{k_1} = c_{k_2}$ whenever k_1 and k_2 belong to the same coset modulo p.

    Moreover, if k_1 and k_2 are in complementary cosets, then c_{k_1} and c_{k_2} are complex conjugates. (The complement of a coset C contains the numbers $p - g$ for all g in C.)

    Proof. If k_1 and k_2 are in the same coset, c_{k_1} and c_{k_2} certainly have the same magnitude, by Equation (14). But, from Equation (13) directly, it is seen that

$$c_k = \frac{1}{p} \sum_m a_{m/k} e^{2\pi i m/p} ,$$

so that c_k is related to c_1 by a reverse decimation of $\{a_j\}$. Two reverse decimations are the same if and only if the corresponding k's are in the same cyclotomic coset, just as in the case of forward decimations. Finally,

$$c_{-k} = \frac{1}{p} \sum_{j=1}^{p} a_j e^{-2\pi i k j/p} = \bar{c}_k ,$$

so that when k_1 and k_2 are in complementary cosets, c_{k_1} and c_{k_2} are complex conjugates. (This theorem can be strengthened to show that $c_{k_1} = c_{k_2}$ if and *only if* k_1 and k_2 belong to the same coset.)

Example. When $p = 15$, the cosets are

$$C_0 = \{0\}$$
$$C_1 = \{1, 2, 4, 8\}$$
$$C_2 = \{7, 14, 13, 11\}$$
$$C_3 = \{3, 6, 12, 9\}$$
$$C_4 = \{5, 10\} .$$

For the sequence 011110101100100, with period 15, the Fourier coefficients, as defined by Equation (13), are

$$c_0 = \frac{8}{15}$$

$$c_1 = c_2 = c_4 = c_8 = -\frac{1}{30} + i\frac{\sqrt{15}}{30}$$

$$c_7 = c_{14} = c_{13} = c_{11} = -\frac{1}{30} - i\frac{\sqrt{15}}{30}$$

$$c_3 = c_6 = c_{12} = c_9 = -\frac{2}{15}$$

$$c_5 = c_{10} = \frac{2}{15} .$$

This sequence $\{c_k\}$ of Fourier coefficients is periodic with period 15. Also, $c_{k_1} = c_{k_2}$ if and only if k_1 and k_2 are in the same coset; whereas $c_{k_1} = \bar{c}_{k_2}$ if k_1 and k_2 are in complementary cosets. (Several cosets are self-complementary; hence the corresponding Fourier coefficients are *real*.) In magnitude,

$$|c_k| = \begin{cases} 8/15 & \text{if } k \text{ is a multiple of 15} \\ 2/15 & \text{for all other } k . \end{cases}$$

It is a simple observation that if $\{a_j\}$ is any periodic sequence whatever, the sum of the Fourier coefficients defined in Equation (13) over one period equals a_0. Specifically,

$$\sum_{k=0}^{p-1} c_k = \frac{1}{p} \sum_{j=1}^{p} a_j \sum_{k=0}^{p-1} e^{2\pi i k j/p} = \frac{1}{p} a_0 p = a_0 . \qquad (16)$$

In the example with $p = 15$,

$$\sum_{k=0}^{p-1} c_k = \frac{8}{15} + 4\left(-\frac{1}{30} + i\frac{\sqrt{15}}{30}\right) + 4\left(-\frac{1}{30} - i\frac{\sqrt{15}}{30}\right)$$

$$+ 4\left(-\frac{2}{15}\right) + 2\left(\frac{2}{15}\right) = 0 = a_0 .$$

If the sequences $\{a_j\}$ and $\{b_j\}$ are time-inverses of each other, their Fourier coefficients differ only by complex conjugation. This is an immediate consequence of Theorem 5.

8. PERFECT SEQUENCES

If $\{a_j\}$ is a periodic binary sequence with period v (where v is any integer), and if the auto-correlation function $C(\tau)$ of $\{a_j\}$ satisfies

$$C(\tau) = \sum_{j=1}^{v} a_j a_{j+\tau} = \begin{cases} k & \text{if } \tau \equiv 0 \pmod{v} \\ \lambda & \text{if } \tau \not\equiv 0 \pmod{v}, \end{cases} \tag{17}$$

then $\{a_j\}$ is called a *perfect sequence*. The crux of the definition is that $C(1) = C(2) = \cdots = C(v-1)$ must hold.

Theorem 6.   Every PN sequence is perfect.

Proof. Let $\{a_j\}$ be a PN sequence with period p. The $\{a_j \oplus a_{j+\tau}\}$ is simply a phase shift of $\{a_j\}$, for any τ, $1 \le \tau \le p-1$, by the cycle-and-add property (see Chapter III for a complete discussion of this property). Since originally $\{a_j\}$ had 2^{n-1} *ones* and $2^{n-1} - 1$ *zeros* in every period, this must still be true of $\{a_j \oplus a_{j+\tau}\}$. Assume that $a_j \oplus a_{j+\tau} = 0$ occurs x times in the form $1 \oplus 1 = 0$. Then the other $2^{n-1} - 1 - x$ times must be of the form $0 + 0 = 0$. There are $2^{n-1} - x$ *ones* in the first term, which must occur as $1 + 0 = 1$, and likewise $2^{n-1} - x$ *ones* in the second term which must occur in the form $0 + 1 = 1$. Altogether there are $p = 2^n - 1$ terms. Hence, $2^n - 1 = x + (2^{n-1} - 1 - x) + (2^{n-1} - x) + (2^{n-1} - x)$, whence $x = 2^{n-2}$, and the number of 1-1 pairs in $a_j a_{j+\tau}$ is 2^{n-2}, independent of τ, $0 < \tau < p$. Therefore

$$C(\tau) = \begin{cases} 2^{n-1} & \text{if } \tau \equiv 0 \pmod{p} \\ 2^{n-2} & \text{if } \tau \not\equiv 0 \pmod{p}, \end{cases}$$

and all PN sequences are perfect.

Only the PN sequences have the cycle-and-add property; but there are many other perfect sequences. Two examples, one of length 11 and one of length 13, are 11011100010 and 1101000001000. The first example has a balance of 1's and 0's, the second does not. Every perfect sequence satisfies

$$k(k-1) = \lambda(v-1) \tag{18}$$

because

$$k^2 = \sum_{\tau=1}^{v} \sum_{j=1}^{v} a_j a_{j+\tau} = \sum_{\tau=1}^{v} C(\tau) = k + (v-1)\lambda . \tag{19}$$

Theorem 7. If $\{a_j\}$ is perfect with period v, and $(q, v) = 1$, then $\{b_j\} = \{a_{qj}\}$ is again perfect.

Proof. Let s be the integer which satisfies $qs \equiv 1$ (modulo v). Then for $\{b_j\}$

$$C_b(\tau) = \sum_{j=1}^{v} b_j b_{j+\tau} = \sum_{j=1}^{v} a_{qj} a_{qj+\tau}$$

$$= \sum_{j=1}^{r} a_{qj} a_{q(j+s\tau)} = \sum_{j=1}^{r} a_j a_{j+s\tau} = C_a(s\tau) = C_a(\tau).$$

Those q which take $\{a_j\}$ back into itself are the multipliers. They form a group which is a subgroup in the multiplicative group of integers modulo v. The perfect sequences $\{a_{q_1 j}\}$ and $\{a_{q_2 j}\}$ are the same if and only if q_1 and q_2 belong to the same coset of the multiplier subgroup. Thus the entire theory of perfect sequences follows the pattern laid out in the special case of PN sequences.

For the PN sequences, the numbers 1, 2, 4, 8, ..., 2^{n-1} were multipliers. More generally, for a (v, k, λ) perfect sequence the prime divisors of $k - \lambda$ are multipliers, and generate a multiplier group. (This theorem is stated in [19], which also contains an excellent summary and bibliography of the (v, k, λ) problem.)

For the sequence 1101000001000, with $v = 13$, $k = 4$, and $\lambda = 1$, the only prime divisor of $k - \lambda$ is 3. Hence 1, 3, and 9 are the multiplier group for this sequence.

Chapter V

ON THE FACTORIZATION OF TRINOMIALS OVER GF(2)

1. INTRODUCTION

There is an extensive literature on the factorization of polynomials over various fields, of which some, notably Selmer's paper [34], concern the factorization of trinomials over the rational field. The factorization of polynomials over finite fields is especially amenable to computation, since for fixed degree there are only finitely many cases to consider. Even more restricted is the study of trinomials, where it is still possible to consider all cases, even when the degree is comparatively large.

Table V-1 contains the irreducible factors of every trinomial over GF(2) with degree ≤ 36, together with the period of each trinomial and of each irreducible factor. In Table V-2, the factor of lowest degree is listed for the trinomials of degrees 37 through 45.

2. THEORETICAL DISCUSSION

We need only consider trinomials of the form $x^n + x^a + 1$ since all others can be reduced to this case by factoring out the appropriate power of x. From divisibility considerations, all factors of such a trinomial will also "end" in 1. A well-known algebraic theorem states that an irreducible polynomial of degree n over GF(2) will always divide $x^{2^n-1} + 1$. We define the least k such that a polynomial divides $x^k + 1$ to be the period of the polynomial. If the polynomial is irreducible and of degree n, then k divides $2^n - 1$ and does not divide $2^i - 1$ $(i < n)$. Thus, if $2^n - 1$ is a Mersenne prime, then all irreducible polynomials of degree n have period $2^n - 1$. (The theory of roots of polynomials over finite fields is treated more fully in [39] and [1] and the special case GF(2) is treated extensively in Chapter III.)

The trinomials $x^n + x^a + 1$ $(n > a)$ are listed in order of increasing n. For each n, they are listed in order of increasing a. Since the transformation $f(x) \rightarrow x^n f(1/x)$ changes any polynomial of degree n into its "reverse," i.e.,

$$\sum_{k=0}^{n} a_k x^k \rightarrow \sum_{k=0}^{n} a_k x^{n-k} ,$$

and does the same for each of its factors, it was sufficient to list only those trinomials with a $\leq [n/2]$. Furthermore, those cases in which both n and a are even are omitted, since $x^{2i} + x^{2j} + 1 = (x^i + x^j + 1)^2$. The periods of the trinomials are listed, followed by the irreducible factors, with their periods, in order of increasing degree. Factors are written in octal notation; i.e., $31107 = 011001001000111 = x^{13} + x^{12} + x^9 + x^6 + x^2 + x + 1$.

It is easy to show that all trinomials listed are square-free. If $x^n + x^a + 1$ had a repeated factor, then its derivative would also be divisible by this factor. But, since n and a are not both even, $nx^{n-1} + ax^{a-1}$ is either a power of x, obviously co-prime to $x^n - x^a + 1$, or of the form $x^{n-1} + x^{a-1}$. But a divisor of both $x^{n-1} + x^{a-1}$ and $x^n + x^a + 1$ also divides $x^n + x^a + 1 + x(x^{n-1} + x^{a-1}) = 1$; so $x^n + x^a + 1$ is proved square-free. The period of such a square-free trinomial is the LCM of the periods of its irreducible factors. (The period of the $2^{k\text{th}}$ power of a trinomial is 2^k times the period of the trinomial.)

3. PROCEDURE

The procedure followed in obtaining the entry for each $x^n + x^a + 1$ is presented below.[1]

1. The exponents n and a were reduced modulo $2^p - 1$ for each p such that $2^p - 1 \leq n$. If the resulting trinomial, of lower degree, was divisible by an irreducible polynomial of period p (which could be determined from previous entries in the table), then this polynomial divided the original trinomial. An alternate method was to test n and a by an equation for each k, such that if the equation is satisfied, a polynomial of period k divides the trinomial. Examples of proper equations are $a + n \equiv 0$ (mod 3) for period 3, and $a^2 - na + n^2 \equiv 0$ (mod 7) for period 7. (Here the cases $n \equiv 0$ (mod 3) and $n \equiv 0$ (mod 7), respectively, must be excluded.) Then, since all irreducible polynomials of period p divide $x^{2^p-1} + 1$, the Euclidean algorithm could be applied to get a factor.

2. After low-degree factors were found by method 1, the remaining polynomial was either irreducible or possessed a factor of degree

---

[1] R. W. Marsh's *Table of Irreducible Polynomials over GF(2) through Degree 19* [23] was of great help.

$\leq (n \ 2)$, where n was its degree.   The Euclidean algorithm, as explained in 1, was then applied for each $p \leq (n,2)$.   For fairly small n, this process eventually gave all factors; otherwise, calculations became too long.   Even for large n, some factors were determined in this way.

   3.   When the testing for factors of every degree became too time-consuming, the period of the trinomial was determined by running an appropriate shift register sequence on a special-purpose digital computer.   This period, the LCM of a finite set of numbers, each of which divides a number of the form $2^p - 1$, was factored by trial and error, thus giving an indication as to the degree and period of each factor. The Euclidean algorithm was then applied to determine factors of the particular periods indicated.

   4.   The above methods sufficed to factor all trinomials of degree ≤ 36, with only a few exceptions.   In several instances, there were several factors of the same degree but of different periods.   In these cases, application of the Euclidean algorithm for all periods dividing $2^n - 1$ (where n is the common degree of the factors) succeeds in separating factors.   (Two factors of the same degree and the same period cannot be so separated.)

   5.   Finally, in some cases, there were several factors of the same degree and the same period.   Unless the degree was small, these presented a real problem.   The best method of attack was to use the transformation group of order 6 generated by $f(x) \to f(x + 1)$ and $f(x) \to x^n f(1 \ x)$ (where n is the degree of f).   This group leaves the *degrees* of factors unchanged, but may change their *periods*.   If this is the case, method 4 may become applicable.

   6.   The "cubic transformation" $f(x) \to f(x^{1/3})f(\rho x^{1/3})f(\rho^2 x^{1/3})$ (cf. [1]), where ρ is a primitive cube root of unity, was also useful in studying the roots and periods of trinomials.   It preserves the periods of polynomials whose periods are not divisible by three, while if the period P is divisible by three, the new period is $P,3$.

   The same methods were used to compile Table 2, which contains the smallest factors of trinomials of degree 37-45, inclusive.

4.   COMPENDIUM OF SPECIAL RESULTS

   Theorem 1.   A trinomial over GF(2) not divisible by x has a repeated factor if and only if the trinomial is a perfect square.   (See Sec. 2 for outline of proof.)

   Theorem 2.   All irreducible factors of $x^{2n} + x + 1$ have degrees dividing $2n$, and therefore, periods dividing $2^{2n} - 1$.

Proof.

$$f(x) = x^{2^n} + x + 1$$

$$g(x) = x^{2^n} f\left(\frac{1}{x}\right) = x^{2^n} + x^{2^n-1} + 1$$

$$h(x) = g(x+1) = \frac{x^{2^{n+1}} + 1}{x+1},$$

so that $h(x)$ divides $x^{2^{n+1}} + 1$, which divides $x^{2^{2n}-1} + 1$. However, the transformation from f to h preserves degrees of factors, so that all factors of f will have degrees dividing $2n$ and periods dividing $2^{2n} - 1$.

Theorem 3. All irreducible factors of $x^{2^{n+1}} + x + 1$ have degrees dividing $3n$, and therefore, periods dividing $2^{3n} - 1$.

Proof.

$$f(x) = x^{2^n+1} + x + 1$$

$$g(x) = xf^{2^n} + f = x^{2^{2n}+2^n+1} + 1$$

so that $f(x)$ divides $g(x)$, which divides $x^{2^{3n}-1} + 1$. Therefore all factors of f have degrees dividing $3n$ and periods dividing $2^{3n} - 1$.

Theorems 1, 2 and 3 are also proved in [11]. The proof of 1, by E. N. Gilbert, was by a completely different method than that above. The proofs given for Theorems 2 and 3 were both somewhat similar to that of 3 above. Gilbert credits 2 and 3 to J. Riordan.

Conjecture. All irreducible factors of $x^{2^n+1} + x^{2^{n-1}-1} + 1$ have degrees dividing $6(n - 1)$ and periods dividing $2^{6(n-1)} - 1$. This has been verified for $n = 1, 2, 3, 4$, and 5. The irreducible factors of $x^{2^n+1} + x^{2^n-2} + 1$ seem to have a pattern similar to Theorems 2, 3, and 4, but a plausible conjecture has not yet been formulated.

Theorem 4. If the set of roots of $f(x)$ is invariant under $x \to 1/x$, then $f(x)$ divides a trinomial only if the roots of $f(x)$ are $3a^{\text{th}}$ roots of unity for some a.

Proof. Assume

$$f(x) | x^n + x^a + 1.$$

Then

$$f(x) = x^n f(1/x) | x^n + x^{n-a} + 1.$$

Therefore

$$f(x) | (x^n + x^a + 1) + (x^n + x^{n-a} + 1) = x^{n-a} + x^a.$$

Therefore

$$f(x)|x^a(x^{n-a} + x^a) + (x^n + x^a + 1) = x^{2a} + x^a + 1.$$

Therefore, the roots of f are $3a^{th}$ roots of unity.

Corollary 1 to Theorem 4. $x^3 + x + 1$ and $x^3 + x^2 + 1$ are never both divisors of a trinomial.

Corollary 2 to Theorem 4. $x^4 + x^3 + x^2 + x + 1$ never divides a trinomial.

Counterexample.   A counterexample has been found to a statement by A. A. Albert.[2]   The trinomial $x^{11} + x^6 + 1$ has an irreducible factor of degree 8 and period 5×17, which furnishes a counterexample to the following statement:

Let $2m = n$, $e = e_1 e_2$, $e_2 | 2^m + 1$, $e_1 | 2^m - 1$. Then for any irreducible polynomial $f(x)$ of degree n and period e, there exists an $\alpha \equiv 0 \pmod{e_1}$ and a β such that $f(x)|x^a + x^\beta + 1$ if and only if 3 divides e.

Theorem 5.   Let

$$f(x) = \sum_{k=1}^{n} a_k x^k$$

and let $f(x)$ be irreducible, with period $2^n - 1$. Then

$$F(x) = \sum_{k=1}^{n} a_k x^{2^k - 1}$$

is irreducible. (If $2^n - 1$ is a Mersenne prime, then $f(x)$ also has maximum period.)

This theorem was first proved by O. Ore [26] and rediscovered by Gleason and Marsh [41].

Theorem 6.   Let $n = 2^i - 1$ and $a = 2^j - 1$, and suppose that the period of $x^n + x^a + 1$ divides $2^k - 1$. Then, for all m, the trinomial with exponents $(2^{mi} - 1)/(2^m - 1)$, $(2^{mj} - 1)/(2^m - 1)$, 0, has a period which is a divisor of $(2^{mk} - 1)/(2^m - 1)$.

This result was first conjectured by J. Riordan[3] and proved for $j < i \le 6$. The general result is a consequence of the following theorem.

Theorem 7.   Let P and Q be polynomials over GF(p), where p is prime, and suppose the only powers of x occurring nontrivially in P and Q are of the form $p^\beta - 1$.

[2] Albert, A. A., *Some Remarks on Trinomial Equations*, University of California at Los Angeles, Los Angeles, California, August 12, 1955 (unpublished).
[3] Unpublished Bell Telephone Laboratory Memorandum, November 18, 1954.

Let D_n be the operator that replaces all exponents of the form $p^\beta - 1$ by $(p^{n\beta} - 1)/(p^n - 1)$. Then $P(x)$ divides $Q(x)$ if and only if $D_nP(x)$ divides $D_nQ(x)$.

Before the proof is given, two lemmas are presented.

Lemma 1. For any n, $P(x)$ divides $Q(x)$ if and only if $P(x^n)$ divides $Q(x^n)$.

The proof is trivial and will not be given.

Definition. The class $\mathscr{E}$ is the class of polynomials meeting the conditions of Theorem 7.

Lemma 2. Let $P, Q \in \mathscr{E}$ and E_n be the operator which replaces the exponents $p^\beta - 1$ by $p^{n\beta} - 1$. Then P divides Q if and only if E_nP divides E_nQ.

Proof. The proof is by complete induction. First, if the degree of $Q(x)$ is less than the degree of $P(x)$, then both divisibility properties are equivalent to the condition that Q is identically zero and thus are equivalent to each other. Next, suppose the degree of Q to be greater than that of P: Let

$$P(x) = \sum_i a_i x^{p^{\alpha_i}-1},$$

where $p^{\alpha_1} - 1$ is the largest exponent, let

$$Q(x) = \sum_i b_i x^{p^{\beta_i}-1},$$

where $p^{\beta_1} - 1$ is the largest exponent, and then form

$$Q_1(x) = Q(x) - \frac{b_1}{a_1} x^{p^{\beta_1-\alpha_1}-1}[P(x)]^{p^{\beta_1-\alpha_1}}. \qquad (1)$$

Since $[F(x)]^p = F(x^p)$ for any polynomial over $GF(p)$,

$$Q_1(x) = Q(x) - \frac{b_1}{a_1} \sum_i a_i x^{p^{\beta_1-\alpha_1+\alpha_i}-1}.$$

Therefore, $Q_1 \in \mathscr{E}$ and Q_1 is of lower degree than Q. Similarly, form

$$Q_2(x) = E_nQ(x) - \frac{b_1}{a_1} x^{p^{n(\beta_1-\alpha_1)-1}}[E_nP(x)]^{p^{n(\beta_1-\alpha_1)}}. \qquad (2)$$

Then

$$Q_2(x) = E_nQ(x) - \frac{b_1}{a_1} \sum_i a_i x^{p^{n(\beta_1-\alpha_1+\alpha_i)-1}}$$

and

$$Q_2(x) = E_nQ_1(x). \qquad (3)$$

Now Eq. (1) implies that $P(x)$ divides $Q(x)$ if and only if $P(x)$ divides $Q_1(x)$, Eq. (2) implies that $E_nP(x)$ divides $E_nQ(x)$ if and only if $E_nP(x)$ divides $Q_2(x)$, and by use of the inductive hypothesis, Eq. (3) implies that $P(x)$ divides $Q_1(x)$ if and only if $E_nP(x)$ divides $Q_2(x)$. This chain of equivalences then implies that $P(x)$ divides $Q(x)$ if and only if $E_nP(x)$ divides $E_nQ(x)$.

Theorem 7 is now an immediate consequence of these lemmas.

Proof. By Lemma 2, $P(x)$ divides $Q(x)$ if and only if $E_nP(x)$ divides $E_nQ(x)$. Now $E_nF(x) = (D_nF)(x^{p^n-1})$, so that by Lemma 1, $E_nP(x)$ divides $E_nQ(x)$ if and only if $D_nP(x)$ divides $D_nQ(x)$. This pair of equivalences implies the theorem.

> *Corollary to Theorem 6.* Riordan's conjecture is an application of Theorem 7 with $p = 2$, $P(x) = x^{2^j-1} + x^{2^i-1} + 1$ and $Q(x) = x^{2^k-1} + 1$.

> *Theorem 8.* If $f(x) = x^n + x^a + 1$ is a trinomial whose period divides p, then the trinomial $F(x) = x^{p-a} + x^{n-a} + 1$ is divisible by $f(x)$.

Proof. Clearly,

$$f(x) = (x^p + 1) + x^a F(x) .$$

Since $f(x)$ divides $x^p + 1$ but not x^a, $f(x)$ must divide $F(x)$.

5. UNSOLVED PROBLEMS[1]

Many basic questions concerning trinomials remain unanswered. For example, no one has proved that there are infinitely many trinomials whose roots have the maximum period. (It is easy to exhibit an infinite class of irreducible trinomials, viz. $x^{2 \cdot 3^a} + x^{3^a} + 1$ for all $a = 0, 1, 2, \ldots$, but whose roots have only 3^{a+1} as their period.)

During the preparation of this trinomial factor table, the purely empirical conjecture arose that there are no irreducible trinomials of degree $8n$, for any value of n. Several mathematicians worked for some considerable time attempting either to prove or disprove this conjecture. Finally, Richard Swan of the University of Chicago obtained this result as a mathematical theorem. More specifically, using deep methods in algebraic number theory, he was able to show that a trinomial of degree $8n$ is always the product of an *even* number of irreducible factors. (In particular, it does not consist of only *one* such factor.)

---

[1] This material was added in assembling these papers for the book.

Table V-1. Trinomial factor table, degrees ≤36.[1]

n	a	Period	Octal₁	Period₁	Octal₂	Period₂	Octal₃	Period₃	Octal₄	Period₄	Octal₅	Period₅
2	1	3	irreducible									
3	1	7	irreducible									
4	1	15	irreducible									
5	1	21	7	3	15	7						
	2	31	irreducible									
6	1	63	irreducible									
	3	9	irreducible									
7	1	127	irreducible									
	2	93	7	3	67	31						
	3	127	irreducible									
8	1	63	7	3	155	63						
	3	217	13	7	57	31						
9	1	73	irreducible									
	2	465	31	15	75	31						
	3	21	15	7	165	21						
10	4	511	irreducible									
	1	889	13	7	271	127						
	3	1023	irreducible									
	5	15	7	3	23	15	31	15				
11	1	1533	7	3	1555	511						
	2	2047	irreducible									
	3	1953	67	31	163	63						
	4	1533	7	3	1553	511						

[1] Several entries in this Table were supplied by Richard W. Marsh.

Table V-1 (Cont'd).

n	a	Period	Octal₁	Period₁	Octal₂	Period₂	Octal₃	Period₃	Octal₄	Period₄	Octal₅	Period₅
12	5	595	13	7	567	85	57	31				
	1	3255	15	7	31	15						
	3	45	irreducible									
	5	819	irreducible									
13	1	7905	73	31	651	255	765	255				
	2	1785	7	3	13	7						
	3	8001	133	63	277	127						
	4	7161	15	7	3515	1023						
	5	6141	7	3	6673	2047						
	6	7665	23	15	1157	511						
14	1	11811	7	3	51	31	345	127				
	3	5115	23	15	2327	1023						
	5	5461	irreducible									
	7	21	7	3	127	21	165	21				
15	1	32767	irreducible									
	2	4599	163	63	1563	511	127	21				
	3	63	13	7	111	9						
	4	32767	irreducible									
	5	35	15	7	16475	35						
	6	93	73	31	3247	93						
	7	32767	irreducible									
16	1	255	551	255	573	85						
	3	57337	15	7	35165	8191						

n	i	value								
17	5	16383	7	3	66673	16383	17523	273	277	127
	7	63457	51	31	5351	2047	3615	1023		
	1	273	7	3	13	7				
	2	114681	irreducible							
	3	131071	7	7	72351	16383	1231	73		
	4	1023	irreducible							
	5	131071	irreducible							
	6	131071	7	3	57	31				
	7	4599	13	3	163	63				
	8	35805	45	7	23	15				
18	1	253921	irreducible							
	3	189	13	31	22637	8191	2475	1023		
	5	32767	irreducible							
	7	262143	irreducible							
	9	27	15	7	134567	32767	67	31		
19	1	413385	7	7	23	15				
	2	129921	31	3	367	127				
	3	491505	13	15	172621	32767	2701	341		
	4	91749	7	7	271347	13107	51	31		
	5	393213	361	3	666673	131071	27221	8191	1207	511
	6	520065	253	127	14661	4095				
	7	520065	7	127	12067	4095				
	8	47523	1175	3	15	7				
	9	174251	7	511	2355	341				
20	1	761763		3	75	31				
	3	1048575	irreducible							

Table V-1 (Cont'd).

n	a	Period	Octal₁	Period₁	Octal₂	Period₂	Octal₃	Period₃	Octal₄	Period₄	Octal₅	Period₅
21	5	75	irreducible									
	7	779907	7	3	357	127	5235	2047				
	9	130305	13	7	543	255	1055	511				
	1	5461	253	127	50331	5461						
	2	2097151	irreducible									
	3	381	323	127	72427	381						
	4	406317	73	31	324427	13107						
	5	5461	217	127	43723	5461						
	6	279	75	31	111	9	3205	93				
	7	49	irreducible									
	8	1966065	31	15	753131	131071						
	9	381	247	127	53373	381						
	10	2088705	435	255	21615	8191						
22	1	4194303	irreducible									
	3	3670009	13	7	2713457	524287	431277	131071				
	5	2752491	7	3	15	7						
	7	4063201	45	31	454765	131071						
	9	3899535	23	15	253	127	5735	2047				
	11	33	7	3	2251	33	3043	33				
23	1	2088705	7	3	515	255	34641	8191				
	2	7864305	23	15	2327423	524287	103653	32767				
	3	32767	15	7	73	31	37371	8191				
	4	2088705	7	3	543	255						

n	i											
24	5	8388607	irreducible									
	6	458745	13	7	31	15	330537	21845				
	7	2094081	7	3	2157	1023	6435	2047				
	8	2728341	141	63	247	127	3133	1023				
	9	8388607	irreducible									
	10	87381	7	3	15	7	1062537	12483	2541	93	13627	35
	11	126945	51	31	133	63	11637	585				
25	1	2097151	13	7	13456771	2097151						
	3	189	111	9	1101101	189						
	5	16766977	5337	2047	24703	8191						
	7	1048575	23	15	4657057	349525						
	9	651	15	7	51	31	165	21				
	11	5586603	2767	1023	55753	16383	7621	2047				
26	1	10961685	147	63	435	255						
	2	25165821	7	3	66666667	8388607						
	3	33554431	irreducible									
	4	2158065	51	31	675	85	15365	4095				
	5	105	7	3	13	7	23	15				
	6	4185601	15	7	1113	73	36403	8191	31	15		
	7	33554431	irreducible									
	8	8322945	7	3	277	127	363255	21845				
	9	32247967	73	31	357	127	24031	8191				
	10	155	57	31	5521623	155	675611	131071				
	11	8257473	7	3	103	63	121563	32767				
	12	4161409	13	7	235	127	1275	511	12515	4095		
	1	298935	7	3	15	7						

Table V-1 (Cont'd).

n	a	Period	Octal$_1$	Period$_1$	Octal$_2$	Period$_2$	Octal$_3$	Period$_3$	Octal$_4$	Period$_4$	Octal$_5$	Period$_5$
	3	2094081	75	31	2773	1023	7173	2047				
	5	67074049	4261	2047	105621	32767	1573411	13797				
	7	13799	7	3	103	63						
	9	7449145	2633	341	272107	21845						
	11	8371713	13	7	1027	73	57201	16383				
	13	39	7	3	13617	39	17075	39				
27	1	125829105	31	15	75310753	8388607						
	2	458745	13	7	717	255	375715	65535				
	3	219	1511	73	1654501	219						
	4	5592405	15	7	164723515	5592405						
	5	8877935	247	127	5337633	69905	11001	45				
	6	1395	45	31	2065	93						
	7	44564395	477	85	2352103	524287						
	8	133693185	607	255	3745133	524287						
	9	63	13	7	127	21	133	63	141	63	147	63
	10	130023393	67	31	34676453	4194303						
	11	109226985	15	7	23	15	313	127	24061	8191		
	12	1533	1665	511	1526345	1533						
	13	130023393	57	31	26507623	4194303						
28	1	17895697	irreducible									
	3	268435455	irreducible									
	5	21082635	7	3	57	31	163	63	323	127	771	85
	7	105	31	15	13321	105	15611	105				
	9	268435455	irreducible									

n	q									
	11	199753347	7	3	217	127	3221475	524287	551461	131071
	13	268435455	irreducible							
29	1	402653181	irreducible							
	2	536870911	13	3	1555555555	134217727	45	31		
	3	426636105	7	7	23	15	1517	511	7723	2047
	4	398532477	15	3	313	127	1541	511	15457	1365
	5	3088995	27765	7	67	31	1437	511	32461	8191
	6	536797185	7	8191	266745	65535				
	7	389260893	3705	3	75	31	301			
	8	178781867	453	341	3065625	524287	253	127	532761	131071
	9	534773505	7	255	11377067	2097151				
	10	349566357	31	3	13	7				
	11	503316465	15	15	365443531	33554431				
	12	469762041	7	7	723517235	67108863				
	13	402653181	1137	3	1555553333	134217727				
	14	178607275	irreducible							
30	1	10845877	15	511	4533443	349525	15125	315		
	3	2667	133	7	165	21			50667	381
	5	315	4671	63	14373	315	23335017			
	7	1073215489	irreducible							
	9	3069	57	2047	2313171	524287				
	11	33554431	13	31	265073623	33554431	11001	4194303		
	13	910163751	111	7	51	31				
	15	45	13	9	10011	45		45		
31	1	2097151	7	7	211	127	13144661	2097151		
	2	22362795		3	15	7	15341	4095	71403	16383

Table V-1 (Cont'd).

n	a	Period	Octal₁	Period₁	Octal₂	Period₂	Octal₃	Period₃	Octal₄	Period₄	Octal₅	Period₅
	3	2147483647	irreducible									
	4	670965765	23	15	27411	8191	52547	5461				
	5	107359437	7	3	20715	8191	337377	4369				
	6	2147483647	irreducible									
	7	2147483647	irreducible									
	8	469762041	7	3	13	7	765142061	67108863				
	9	24956085	15	7	717	255	4203427	209715				
	10	712248405	765	255	3753	341	24513	8191				
	11	536606721	7	3	7363	2047	1247631	262143				
	12	94371795	31	15	147	63	12747223	2097151				
	13	2147483647	irreducible									
	14	536821761	7	3	53545	5461	165547	32767				
	15	2097151	13	7	203	127	13535415	2097151				
32	1	1023	7	3	3417	341	3435	1023	3543	1023		
	3	1409286123	103	63	414247507	67108863						
	5	3758096377	13	7	5627134567	536870911						
	7	97612893	7	3	15555555533	97612893						
	9	268435455	31	15	3654436531	268435455						
	11	1431562923	52621	16383	1206221	262143						
	13	910163751	7	3	15	7	73	31	33743675	4194303		
	15	4292868097	5205	2047	12133725	2097151						
33	1	1057	15	7	103437	1057	166311	1057				
	2	408944445	16571	1365	15123335	2097151	1252621	1533				
	3	4599	111	9	1321	511						

4	2348810205	13	7	31	15	661277557	22369621				
5	1186461481	45	31	1231	73	2764355	524287				
6	6141	6525	2047	35554505	6141	12077651	299593				
7	9287383	67	31	323	127	176753	32767				
8	32767	15	7	117143	32767	1110011	189	435	255	703	255
9	5859	75	31	3205	93						
10	1227133513	irreducible									
11	77	13	7	13456271627	77	1071605	1533				
12	4599	111	9	1175	511						
13	8589934591	irreducible									
14	383290515	23	15	5733	2047	1255515	37449				
15	1785	15	7	165	21	643	85				
16	2796549525	103	63	271	127	5561027	349525	163767	32767	701361	
34 1	255652815	23	15	11536115361	51130563	122773761	2396745				
3	16777215	31	15	141	63						
5	19168695	7	3	51	31	10175	4095				
7	5726623061	irreducible									
9	15032385529	13	7	27134562627	2147483647						
11	10836557067	7	3	15	7	67	31	211	127		
13	17179312129	113625	32767	2257305	524287	2025373	524287				
15	2411195913	133	63	1365	511	661	51	717	51	747	51
17	51	7	3	433	51						
35 1	3681400539	7	3	15555555555	1227133513						
2	3435973837	irreducible									
3	34225520385	561	255	1307503771	134217727						
4	2818056213	7	3	133	63	25333	8191	60367	16383	131071	

Table V-1 (Cont'd).

n	a	Period	$Octal_1$	$Period_1$	$Octal_2$	$Period_2$	$Octal_3$	$Period_3$	$Octal_4$	$Period_4$	$Octal_5$	$Period_5$
	5	635	367	127	3121607613	635						
	6	1108343775	142641	32767	7664741	1048575						
	7	147	7	3	127	21	165	21	10040001	49		
	8	3435945153	547411	131071	1352111	262143						
	9	11408506795	643	85	1640514337	134217727	31	15	45	31	4454725	155
	10	465	7	3	12624165	15						
	11	270532479	51445	16383	2097151							
	12	1632982333	1511	73	726240411	22369621						
	13	2815798293	7	3	103	63	3471	1023	500061	131071	11001	45
	14	217	67	31	120247	217	151265	217				
	15	635	323	127	3521343157	635						
	16	25769803773	7	3	155555533333	3589934591						
	17	119304647	57	31	13243567623	119304647						
36	1	2283952821	1243	511	2257	341	540663	131071				
	3	9765	13	7	51	31	127	21				
	5	59894659335	15	7	515	255	332176461	33554431	2541	93		
	7	4403459697	73	31	361	127	132121467	1118481				
	9	135	irreducible									
	11	68719476735	irreducible									
	13	65535	23	15	323143	65535	372705	21845				
	15	2457	irreducible									
	17	1224736695	13	7	1151	511	122510717	16777215				

Table V-2. Trinomial factor table, degrees 37 through 45.[1]

n	a	Octal	n	a	Octal	n	a	Octal	n	a	Octal
37	1	1543741		9	1011011		16	7		21	23
	2	7		10	477		17	155	44	1	7
	3	15		11	73		18	15		3	15
	4	147		12	13		19	7		5	irreducible
	5	7		13	15		20	maximal		7	7
	6	13		14	maximal	42	1	31		9	211
	7	12705		15	111		3	111		11	31
	8	7		16	3411757		5	331		13	7
	9	23		17	133		7	irreducible[2]		15	155
	10	15		18	10011		9	10011		17	15
	11	7		19	13		11	23		19	7
	12	4653	40	1	15		13	reducible		21	reducible
	13	13		3	567		15	66313	45	1	13
	14	7		5	7		17	reducible		2	15
	15	235		7	57		19	45		3	152323
	16	57		9	7300371		21	103		4	67
	17	7		11	7	43	1	57		5	1145
	18	2443		13	376475		2	7		6	1110011
38	1	7		15	15		3	13		7	reducible
	3	235		17	7		4	141		8	13
	5	73		19	221		5	7		9	15
	7	7	41	1	7		6	23		10	23
	9	15		2	13		7	31		11	reducible
	11	42645		3	maximal		8	7		12	152345
	13	7		4	7		9	reducible		13	45
	15	13		5	551		10	13		14	reducible
	17	23		6	51		11	7		15	13
	19	7		7	7		12	15		16	15
39	1	163		8	253		13	67		17	reducible
	2	1055		9	13		14	7		18	57
	3	57		10	7		15	reducible		19	reducible
	4	maximal		11	15		16	reducible		20	31
	5	13		12	23		17	7		21	123433
	6	15		13	7		18	reducible		22	13
	7	23		14	31		19	15			
	8	maximal		15	3531		20	7			

[1] Several entries in this Table were supplied by Richard W. Marsh.
[2] Non-maximal.

If there are systematic families of degrees other than the multiples of 8 for which no irreducible trinomials exist, this has not been demonstrated. Also, the study of trinomials over finite fields other than GF(2) has not been undertaken with comparable thoroughness.

PART THREE

THE NONLINEAR THEORY

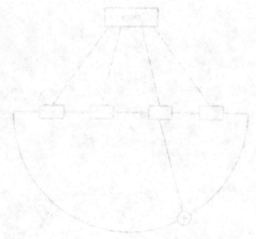

Chapter VI

NONLINEAR SHIFT REGISTER SEQUENCES

1. INTRODUCTION

An n-stage shift register is a device consisting of n consecutive 2-state memory units (usually "flip-flops" or delay positions) regulated by a single clock. At each clock pulse, the state (1 or 0) of each memory stage is shifted to the next stage in line. A shift register is converted into a code generator by including a feedback loop, which computes a new term for the first stage, based on the n previous terms.

In Figure VI-1, a shift register uses the formula $a_k = a_{k-1} \oplus a_{k-4}$ to compute the new terms of the sequence it generates. Here, the symbol $\oplus$ designates addition modulo 2, which obeys the laws $0 \oplus 0 = 1 \oplus 1 = 0$, $0 \oplus 1 = 1 \oplus 0 = 1$. If the initial state in Figure VI-1 is 1111, then the sequence of successive shift register states is as shown in Table VI-1.

Since the sixteenth state in Table VI-1 is the same as the first

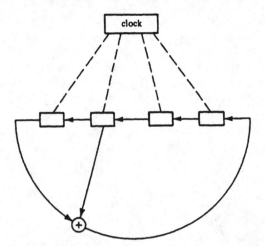

Fig. VI-1. A simple shift register with feedback.

110

Table VI-1. Sequence of successive shift register states corresponding to the feedback formula $a_k = a_{k-3} \oplus a_{k-4}$.

1.	1111	9.	0011
2.	1110	10.	0110
3.	1100	11.	1101
4.	1000	12.	1010
5.	0001	13.	0101
6.	0010	14.	1011
7.	0100	15.	0111
8.	1001	16.	1111

state, and since each state completely determines the next state, this
shift register sequence is periodic, with period $p = 15 = 2^4 - 1$. All
possible binary vectors of length 4, except 0000, occur in this shift
register sequence. It is clearly impossible to obtain a sequence longer
than 2^n from an n-stage shift register.

It is well known that sequences of length $2^n - 1$ can always be
obtained from an n-stage shift register by means of a feedback logic
consisting entirely of modulo-2 additions. The number of modulo-2
logics (or *linear logics*, as they are also called) yielding the maximum
length of $2^n - 1$ is known to be exactly $\phi(2^n - 1)/n$. Here, ϕ is Euler's
function, which is discussed more fully in Section 2.1. To a first approximation, the number of linear logics yielding maximum length is
$2^n/n$. Sometimes a linear logic for maximum length involves only 2
taps (as in the example of Fig. VI-1). These linear 2-tap configura-
tions have been tabulated through degree 35 in Ch. V, Table V-1, and
all the modulo-2 shift registers yielding maximum length have been
determined elsewhere [23] through degree 19.

Removal of the restriction that the feedback logic be linear in-
creases the number of maximum-length shift register codes of degree
n from less than $2^n/n$ to exactly $2^{2^{n-1}}/2^n$. (This formula, discovered
by de Bruijn, is discussed at greater length in Section 2.2.) This
astronomical increase in the number of good codes justifies, in itself,
the quest for practical nonlinear shift registers. The equivalence of
the linear codes to *delay-and-add* sequences (cf. p. 75) serves as a further
impetus to investigation of the nonlinear case.

2. MATHEMATICAL TREATMENT

In this section, the correlation properties and cycle lengths of shift
register sequences are investigated from a theoretical standpoint, and
methods for obtaining all maximum-length sequences are presented.

2.1. *General Theory*

Several theorems of a very general character, concerning both the cycle lengths and correlation properties of shift register sequences, are now derived.

Correlation and run properties. The auto-correlation of a periodic sequence $\{a_k\}$ of period p is defined as

$$C(\tau) = \sum_{k=1}^{p} a_k a_{k+\tau} .$$

(Some authors include a normalizing factor of $1/p$.)

The binary sequence $\{a_k\}$ is said to have *two-level correlation* if $C(0) = \kappa$, and $C(\tau) = \lambda$ for all $\tau \neq 0$ (modulo p). The two-level property is independent of which two real numbers occur in $\{a_k\}$. For present purposes, $\{a_k\}$ is a sequence of 1's and 0's, though reference is made subsequently to $b_k = 2a_k - 1$, the corresponding sequence of 1's and -1's.

First,

$$\sum_{\tau=1}^{p} C(\tau) = \kappa^2 ,$$

because

$$\sum_{\tau=1}^{p} C(\tau) = \sum_{\tau=1}^{p} \sum_{k=1}^{p} a_k a_{k+\tau}$$

$$= \sum_{k=1}^{p} a_k \sum_{\tau=1}^{p} a_{k+\tau}$$

$$= \sum_{k=1}^{p} a_k \kappa$$

$$= \kappa^2 .$$

Second, for all two-level correlation sequences,

$$\kappa(\kappa - 1) = \lambda(p - 1) ,$$

because

$$\kappa^2 = \sum_{\tau=1}^{p} C(\tau)$$

$$= \sum_{\tau=1}^{p-1} C(\tau) + C(p)$$

$$= \lambda(p - 1) + \kappa .$$

Third, if $\{a_k\}$ has the two-level property, so also has $\{a_{qk}\}$, for any q relatively prime to p.

The maximum-length linear shift register sequences of degree n

(length $2^n - 1$) satisfy the two-level correlation condition, with $p = 2^n - 1$, $\kappa = 2^{n-1}$, and $\lambda = 2^{n-2}$. As q runs through all $\phi(2^n - 1)$ values relatively prime to $2^n - 1$ from 1 to $2^n - 1$, the sequence $\{a_{qk}\}$ becomes each of the $\phi(2^n - 1)/n$ maximum-length linear sequences exactly n times. (Two values of q correspond to the same sequence if and only if they differ by a factor of 2^m modulo $2^n - 1$.)

If the maximum-length linear shift register sequence $\{a_k\}$ is replaced by the corresponding sequence $\{b_k\}$ of 1's and -1's, then

$$C(0) = 2^n - 1$$
$$C(\tau) = -1 \qquad (0 < \tau < 2^n).$$

If a shift register sequence has period 2^n, then every possible binary vector of length n must occur exactly once in each period. In particular, a run of n 1's, as well as a run of n 0's, occurs exactly once. Furthermore, a run of $n - 2$ ones, as well as a run of $n - 2$ zeros, occurs exactly once. There are 2 runs of 1's (and 2 runs of 0's) of length $n - 3$, and, in general, there are 2^i runs of 1's (and 2^i runs of 0's) of length $n - i - 2$, for $0 \leq i \leq n - 3$. The maximum-length linear shift register sequences, which have length $2^n - 1$, have exactly this normal run distribution, except that the longest run of 0's has length $n - 1$ instead of length n.

The number of binary sequences of period 2^n with the normal run distribution mentioned above is

$$\frac{1}{2^{n-2}}\left\{\frac{(2^{n-2})!}{\prod_{i=3}^{n}(2^{n-i})!}\right\}^2 \simeq \frac{2^{2^n-(n-3)(n-2)/2}}{16C(2\pi)^{n-3}},$$

where $e^2 < C < e^{7/3}$, which is slightly larger than the number $2^{2^{n-1}-n}$ of shift register sequences of degree n with period 2^n, for all $n > 2$.

The normal run distribution specifies the correlation function $C(\tau)$ for $\tau = 0, \pm 1, \pm 2$, but no further. The shift register sequences of period 2^n, however, have standard correlation functions for all τ with $|\tau| < n$. This fact is expressed quantitatively in Theorem 10.

Basic concepts and examples. An n-stage shift register is properly described as a device which takes any input vector $(a_1, a_2, \ldots, a_n)$ and computes an output vector $(a_2, a_3, \ldots, a_{n+1})$. Thus, a shift register is completely specified by the table of values for a_{n+1}, which it computes for all possible input vectors $(a_1, a_2, \ldots, a_n)$. This table is called the *truth table of the feedback logic*, or, simply, the *truth table of the shift register*. An example of such a truth table for a 3-stage register is given in Table VI-2.

An equivalent representation of the shift register shows, by means

Table VI-2. Truth table for the shift register whose output is the sequence...00011101...repeating with period 8.

	Input		Output
a_1	a_2	a_3	a_4
0	0	0	1
0	0	1	1
0	1	0	0
0	1	1	1
1	0	0	0
1	0	1	0
1	1	0	1
1	1	1	0

of arrows, how successors are assigned by the shift register to the possible input vectors. This representation, called the *vector diagram*, is illustrated for the register of Table VI-2 in Figure VI-2.

Mathematically, then, a shift register is any mapping T of binary n-space Ω_n into itself which satisfies the relation

$$T(a_1, a_2, \ldots, a_n) = (a_2, a_3, \ldots, a_{n+1}) .$$

Each vector of Ω_n is assigned a unique successor by the mapping T. An illustration of the possible irregularity of such mappings is given by the feedback formula $a_n = a_{n-2}a_{n-3}$. The truth table for this shift

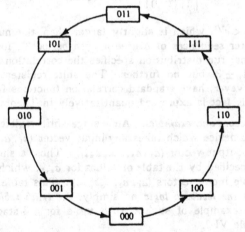

Fig. VI-2. Vector diagram for the shift register of Table VI-2.

Table VI-3. Truth table for the shift register using the feedback logic $a_k = a_{k-2}a_{k-3}$.

Input			Output
a_1	a_2	a_3	a_4
0	0	0	0
0	0	1	0
0	1	0	0
0	1	1	0
1	0	0	0
1	0	1	0
1	1	0	1
1	1	1	1

register is shown in Table VI-3, and the corresponding vector diagram appears in Figure VI-3. This is in sharp contrast to Figure VI-2, as well as to the vector diagram (Figure VI-4) for the shift register which obeys $a_k = a_{k-1} \oplus a_{k-2} \oplus a_{k-3} \oplus a_{k-4}$.

Two valuable theorems. The difference between Figure VI-3, on the one hand, and Figures VI-2 and VI-4, on the other, can be expressed in many equivalent ways. This is the substance of Theorem 1, wherein the statements hold for Figures VI-2 and VI-4, but fail for Figure VI-3.

Theorem 1. The following nine conditions are equivalent:

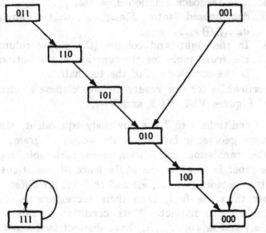

Fig. VI-3. Vector diagram for the shift register of Table VI-3.

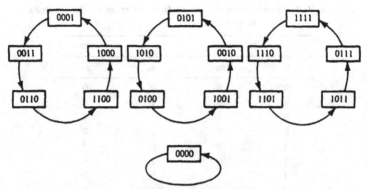

Fig. VI-4. Vector diagram for the shift register using the feedback logic $a_k = a_{k-1} \oplus a_{k-2} \oplus a_{k-3} \oplus a_{k-4}$.

1. The cycles in the vector diagram have no branch points.
2. Every vector has a predecessor as well as a successor.
3. Predecessors of vectors, when they exist, are unique.
4. Distinct vectors have distinct successors.
5. The mapping of Ω_n into Ω_n is *one-to-one*.
6. The mapping of Ω_n into Ω_n is *onto*.
7. The mapping of Ω_n into Ω_n is a permutation of Ω_n.
8. The feedback relation $a_k = F(a_{k-1}, \ldots, a_{k-n})$ can be decomposed into $F(a_{k-1}, \ldots, a_{k-n}) = F_1(a_{k-1}, \ldots, a_{k-n+1}) \oplus a_{k-n}$.
9. In the right-hand column (the *output* column) of the truth table for the function F, the bottom half is the complement of the top half.

It is worthwhile for the reader to check these conditions for the examples of Figures VI-2, VI-3, and VI-4.

Proof. Conditions 1 to 7 are obviously equivalent, since they express the same geometric fact about the vector diagram. Condition 9 is merely the translation of condition 8 into truth-table language. The crux of the proof is to show the equivalence of conditions 4 and 8.

If the two vectors $(\alpha_1, \ldots, \alpha_n)$ and $(\beta_1, \ldots, \beta_n)$ differ in any component other than the first, then their successors $(\alpha_2, \ldots, \alpha_{n+1})$ and $(\beta_2, \ldots, \beta_{n+1})$ are still distinct. Thus, condition 4 holds if and only if $(\alpha_1, \alpha_2, \ldots, \alpha_n)$ and $(\alpha_1', \alpha_2, \ldots, \alpha_n)$ have distinct successors. Here, the prime denotes complementation.

This simply says that

$$F(a'_1, \alpha_2, \ldots, \alpha_n) = F'(\alpha_1, \alpha_2, \ldots, a_n) .$$

Defining $F_1(\alpha_2, \ldots, \alpha_n) = F(0, \alpha_2, \ldots, \alpha_n)$, the previous expression may be written

$$F(\alpha_1, \alpha_2, \ldots, \alpha_n) = F_1(\alpha_2, \ldots, \alpha_n) \oplus \alpha_1 .$$

But this is condition 8, except for a minor change in subscript notation.

In the ensuing discussion all shift registers are assumed to satisfy the conditions of Theorem 1, unless otherwise indicated. There is an obvious loss of sequence length if the cycles of the vector diagrams are allowed to have branches, as in Figure VI-3. This loss is serious if the feedback logic uses only a few of the n available tap positions, which will be the case in most practical code-generating schemes.

Theorem 2. Let $C(\tau)$ be the auto-correlation $\sum_{k=1}^{p} b_k b_{k+\tau}$ corresponding to the shift register sequences $\{a_k\}$ of period p, with $b_k = 2a_k - 1$. As $p \to 2^n$, then $C(\tau) \to 0$ for $\tau = 1$, $2, \ldots, n-1$. However, $C(n) \to 0$ if and only if the feedback logic F_1 (see Theorem 1, condition 8) has an equal number of 1's and 0's in its truth table. (Equivalently, the function F must have equal numbers of 1's and 0's in the top half of its truth table and, consequently, in the bottom half as well.)

Proof. The number p represents not only the period of the sequence, but also the number of vectors in the appropriate cycle of the vector diagram. If p is close to 2^n, then almost all possible vectors occur in the cycle. In the set of all 2^n binary vectors of length n, the first component agrees with the ith component exactly as often as it disagrees, for $\tau = i - 1 = 1, 2, \ldots, n-1$. This is the first part of Theorem 2.

For the case $\tau = n$, one may write $a_{n+1} = F_1(a_n, \ldots, a_2) \oplus a_1$, by virtue of Theorem 1. Thus, $a_{n+1} = a_1$ whenever $F_1 = 0$, whereas $a_{n+1} \neq a_1$ whenever $F_1 = 1$. That these two situations should occur with equal frequency is equivalent, on the one hand, to $\lim_{p \to 2^n} C(n) = 0$ and, on the other hand, to a *balanced truth table* for F_1.

If a genuinely nonlinear feedback logic is to satisfy both the branches-without-cycles condition of Theorem 1 and the balanced-truth-table condition of Theorem 2, it must involve at least 4 tap positions. Let $w = a_{k-n}$, and let x, y, z be three distinct choices from the set of tap positions $a_{k-1}, a_{k-2}, \ldots, a_{k-n+1}$. Then, the genuinely nonlinear 4-

tap logics satisfying Theorems 1 and 2 are as follows:

1. $w \oplus xy \oplus z$
2. $w \oplus x'y \oplus z$
3. $w \oplus xy \oplus x'z$
4. $w \oplus x'y \oplus y'z$
5. $w \oplus xy \oplus z'$
6. $w \oplus xy \oplus y'z'$
7. $w \oplus xy \oplus xz \oplus yz$
8. $w \oplus x'y \oplus x'z \oplus yz$.

Two logics which differ only by permutation of x, y, z, or by simultaneous complementation of x, y, z, have not been regarded as distinct. If individual complementation of x, y, and z is also allowed, the list reduces to three logics, represented by:

i. $w \oplus xy \oplus xz \oplus yz$
ii. $w \oplus xy \oplus y'z$
iii. $w \oplus xy \oplus z$.

Altogether there are $\binom{8}{4} = 70$ ways of writing a truth table for F_1 with 1's in 4 of the 8 positions; the list of eight logics presented above corresponds to removal of the linear logics, the degenerate cases, and the symmetries of permutation and complementation.

A combinatorial problem. The simplest possible shift register of degree n is the one which merely cyclically permutes a sequence of n 1's and 0's. The feedback logic here is

$$a_k = a_{k-n}$$

which satisfies Theorem 1 (condition 8) for producing branchless cycles. In fact, in this case, $F_1(a_{k-1}, \ldots, a_{k-n+1}) \equiv 0$. Condition 9 of Theorem 1 is also satisfied in a particularly simple form. Consider this *cycling register* for $n = 3$. The truth table appears as Table VI-4. The top half of the *output* column (the a_k column) is entirely 0's, whereas the bottom half is entirely 1's. This clearly holds for all cycling registers, which, therefore, do not satisfy the balanced-truth-table condition of Theorem 2. The vector diagram corresponding to Table VI-4 is shown in Figure VI-5.

The purpose of this discussion is to derive the formula for the number $Z(n)$ of cycles into which binary n-space is decomposed by the cycling n-register and to prove that this number is always even, ex-

Table VI-4. Truth table for the cycling register $a_k = a_{k-3}$.

Input			Output
a_{k-3}	a_{k-2}	a_{k-1}	a_k
0	0	0	0
0	0	1	0
0	1	0	0
0	1	1	0
1	0	0	1
1	0	1	1
1	1	0	1
1	1	1	1

cept in the rather trivial case $n = 2$.

The Euler function ϕ may be defined in any of the following ways:

1. $\phi(d)$ is the number of terms t in the sequence $1, 2, \ldots, d$ which satisfy $(d, t) = 1$. The notation (d, t) denotes the greatest common divisor of d and t, whence $(d, t) = 1$ means that d and t are coprime (have no common factors except 1).

2. $\phi(d) = d \prod_{p_i|d} (1 - p_i^{-1})$, the product being extended over all distinct prime divisors p_i of d.

3. $\phi(d) = \prod p_i^{s_i-1} (p_i - 1)$, where $n = \prod p_i^{s_i}$.

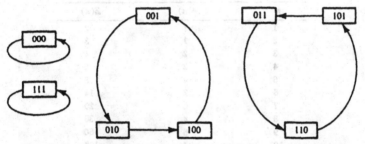

Fig. VI-5. The four cycles into which binary 3-space is decomposed by the cycling register $a_k = a_{k-3}$.

The second and third definitions are shown to follow from the first. It is convenient to rely on the following elementary rule of counting. In a collection of D objects, suppose that: D_1 have the property p_1; D_2 have the property p_2, ...; D_{12} have both properties p_1 and p_2, ...; D_{123} have properties p_1, p_2, and p_3; and so forth. Then, the number of objects having none of the properties $p_1, p_2, \ldots$ is exactly

$$D - D_1 - D_2 - \cdots + D_{12} + D_{13} + \cdots - D_{123} - \cdots .$$

Let the objects be the numbers $1, 2, \ldots, d$. Let the divisors of d be $p_1, p_2, \ldots, p_k$. Let property p_i refer to divisibility by p_i. The rule stated above indicates that the number of objects in the collection which are divisible by none of the prime divisors of d (and, hence, prime to d) is exactly

$$d - \frac{d}{p_1} - \frac{d}{p_2} - \cdots - \frac{d}{p_k} + \frac{d}{p_1 p_2} + \frac{d}{p_1 p_3}$$

$$+ \cdots - \frac{d}{p_1 p_2 p_3} \cdots = d\left(1 - \frac{1}{p_1}\right)\left(1 - \frac{1}{p_2}\right)\cdots\left(1 - \frac{1}{p_k}\right).$$

Thus, the second definition follows from the first.

Using $d = \prod p_i^{s_i}$ and $\phi(d) = d \prod (1 - p_i^{-1})$, one obtains

$$\phi(d) = \prod p_i^{s_i} (1 - p_i^{-1})$$

$$= \prod p_i^{s_i - 1} (p_i - 1),$$

so that the third definition follows from the second. Values of $\phi(n)$ for small n, as well as values of $Z(n)$, appear in Table VI-5.

Table VI-5.   Values of Euler's function $\phi(n)$ and of the cycle function $Z(n)$ for $1 \leq n \leq 12$.

n	$\phi(n)$	$Z(n)$
1	1	2
2	1	3
3	2	4
4	2	6
5	4	8
6	2	14
7	6	20
8	4	36
9	6	60
10	4	108
11	10	188
12	4	352

Theorem 3. The number of cycles $Z(n)$ is given by the equation

$$Z(n) = \frac{1}{n} \sum_{d|n} \phi(d) 2^{n/d}$$

where the summation is extended over all divisors d of n.

Proof. A general formula for the number of cycles or equivalence classes into which a space S is decomposed by a finite group G of operators is

$$\frac{1}{\gamma} \sum_{g \in G} I(g),$$

where γ is the order of the group G (i.e., the number of operators), and $I(g)$ is the number of points of S left fixed by the operator g of G.

In the case at hand, S is binary n-space, with 2^n elements, and G consists of n cyclic permutations of the components, which may be indexed by $i = 1, \ldots, n$. It is easy to obtain $I(g_i) = 2^{(n,i)}$. Setting $(n, i) = d$, one obtains

$$Z(n) = \frac{1}{n} \sum_{i=1}^{n} 2^{(n,i)}$$

$$= \frac{1}{n} \sum_{d \mid n} \sum_{(n,i)=d} 2^d$$

$$= \frac{1}{n} \sum_{d \mid n} 2^d \sum_{(n/d,i)=1} 1$$

$$= \frac{1}{n} \sum_{d \mid n} \phi\left(\frac{n}{d}\right) 2^d$$

$$= \frac{1}{n} \sum_{d \mid n} \phi(d) 2^{n/d}.$$

The last equality is merely a reversal in the order of summation.

Theorem 4. The cycle number $Z(n)$ is even for all $n \neq 2$.

Proof. It is convenient to distinguish three cases:

Case 1. If n is odd, $1/n$ does not disturb the evenness of $\sum \phi(d) 2^{n/d}$, which is even because all the powers of 2 which occur are even.

Case 2. Suppose that $n = 2^k$, with $k \geq 2$. Then,

$$Z(n) = \frac{1}{2^k} \sum_{i=0}^{k} \phi(2^i) 2^{2^{k-i}}$$

$$= 2^{2^k-k} + \sum_{i=1}^{k} 2^{-k+(i-1)+2^{k-i}}.$$

Those terms are even for which $f(k, i) = 2^{k-i} - (k - i) - 1$ is positive. The exceptional terms are $f(k, k - 1) = f(k, k) = 0$. However, the presence of two odd terms keeps the sum even. Only $k = 1$ is not covered, since $i = k - 1 = 0$ corresponds to the even term 2^{2^k-k}; and, indeed, $Z(2)$ is odd.

Case 3. The remaining values of n can be written as $n = 2^k m$, with $k > 0$ and odd $m > 1$. Here,

$$Z(n) = \frac{1}{2^k m} \sum_{m_j | m} \sum_{i=0}^{k} \phi(2^i m_j) 2^{2^{k-i}(m/m_j)}$$

$$= \frac{1}{m} \sum_{m_j | m} \phi(m_j) \frac{1}{2^k} \sum_{i=0}^{k} \phi(2^i) 2^{2^{k-i}(m/m_j)} .$$

The factor $1/m$ does not affect parity. The factor $\phi(m_j)$ is even for $m_j > 1$. The factor $2^{-k} \sum_{i=0}^{k} \phi(2^i) 2^{2^{k-i}(m/m_j)}$, by comparison with case 2, is always an integer, and is even if $m_j = 1$, because then $m/m_j > 1$. Thus, $Z(n)$ is entirely a sum of even terms.

Parity of the cycle number. The following result applies to all shift registers which satisfy the branchless-cycles condition of Theorem 1.

> *Theorem 5.* For $n > 2$, the number of cycles into which binary n-space Ω_n is decomposed by a shift register is even or odd according to whether the number of 1's in the truth table of $F_1(a_{k-1}, \ldots, a_{k-n+1})$ is even or odd.

Proof. By Theorem 1, the complete feedback logic is $F = F_1 \oplus a_{k-n}$. The case $F_1 \equiv 0$ is settled by Theorem 4. Suppose that the number of cycles for a given F_1 is known, and that a single entry in the truth table for F_1 is complemented. By Theorem 1 (condition 9), a certain entry in the *bottom* half of the truth table of F must also be complemented.

A changed entry in the truth table corresponds to a rerouted arrow in the vector diagram. The change in F_1 means that some vector is assigned a new successor. There are two cases: either the new successor is on the same cycle, or it is on a different cycle.

In the case of the same cycle, suppose that the vector V_0 formerly had V_1 as its successor, but now has V_1^* (see Fig. VI-6). Then the old successor V_1 now has no predecessor, and the old predecessor of V_1^* (called V_0^*) now has no successor. The single change in the bottom half of the truth table of F retains the validity of Theorem 1 and, hence, must assign V_1 as the new successor of V_0^*. Thus, in the case of the same cycle, the number of cycles is *increased by one cycle*.

If the new successor V_1^* lies on a new cycle (Fig. VI-7), then, as before, the compensating change in the bottom of the truth table of F must assign to V_0^* (the old predecessor of V_1^*) the new successor V_1. This has the effect of fusing two cycles (the cycles originally containing V_0 and V_1^*, respectively) into one. Thus, in the case of the new cycle, the number of cycles is *decreased by one cycle*.

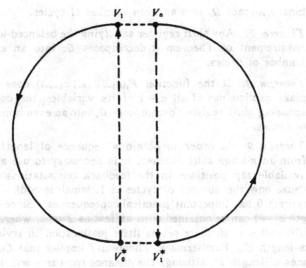

Fig. VI-6. Effect of changing the truth table: case of the same cycle.

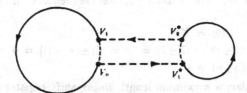

Fig. VI-7. Effect of changing the truth table: case of a new cycle.

In any case, then, a single change in the truth table of F_1 reverses the parity of the number of cycles. Starting from Theorem 4 (the case $F_1 \equiv 0$), the proof of Theorem 5 is completed.

An alternative to using Theorem 4 is to utilize the fact that there is always a two-cycle case (respective lengths $2^n - 1$ and 1) with a purely linear logic. Deforming this cycle pattern t times will thus affect the parity of the cycle number by the parity of t. With linear logic, F_1 has equally many 1's and 0's, which implies an even number of 1's if $n > 2$, but an odd number if $n = 2$. This suffices to prove Theorem 5, but is objectionable on the grounds that it uses deeper and more specialized knowledge to prove a very general result.

The next five theorems are direct corollaries of Theorem 5. It is assumed, for convenience, that $n > 2$, throughout.

Theorem 6. A linear shift register always decomposes

binary n-space Ω_n into an even number of cycles.

Theorem 7.   Any shift register satisfying the balanced-logic requirement of Theorem 2 decomposes Ω_n into an even number of cycles.

Theorem 8.   If the function $F_1(a_{k-1}, \ldots, a_{k-n+1})$ does not make explicit use of all $n - 1$ of its variables, the corresponding shift register decomposes Ω_n into an even number of cycles.

Theorem 9.   In order to obtain a sequence of length 2^n from an n-stage shift register, it is necessary to use all n available tap positions in the feedback computations (because now the number of cycles is 1, which is odd).

Theorem 9 has important practical consequences. Since sequences of length $2^n - 1$ can be obtained with as few as 2 taps, whereas length 2^n requires all n taps, there seems little justification in trying to mechanize length 2^n. Furthermore, Theorem 7 implies that $C(n) \neq 0$ for sequences of length 2^n, although the distance from zero will frequently be small. This implication is now expressed formally.

Theorem 10.   For a shift register sequence of maximum length $p = 2^n$,

$$C(0) = 2^n$$
$$C(\pm 1) = C(\pm 2) = \cdots = C[\pm(n - 1)] = 0$$
$$C(\pm n) \neq 0 .$$

In particular, a maximum-length *linear* shift register sequence can be increased in length from $2^n - 1$ to 2^n by means of the feedback logic $F_{new} = F_{old} \oplus a'_{k-1} a'_{k-2} \cdots a'_{k-n+1}$. In this case, $C(0) = 2^n$, $C(\pm 1) = \cdots = C(\pm(n - 1)) = 0$, and $C(\pm n) = -4$.

Families of shift register sequences.   The unrestricted shift registers of n stages number exactly 2^{2^n}, which is the number of ways of filling in a truth table with 2^n entries. If Theorem 1 is to hold, the number of independent entries reduces to 2^{n-1}, whence there are $2^{2^{n-1}}$ such shift registers. From here, further specialization in two mutually exclusive directions is possible. The lattice hierarchy appears in Figure VI-8.

On the one hand, shift registers may be sought which place all 2^n binary n-vectors on a single cycle. There are $2^{2^{n-1}-n}$ of these, by a result of de Bruijn which is discussed in Section 2.2. On the other hand, the balanced-logic condition of Theorem 2 may be imposed. There are $\binom{2^{n-1}}{2^{n-2}}$ shift registers of this sort, none of them yielding only a single cycle (see Theorem 7). By Stirling's formula,

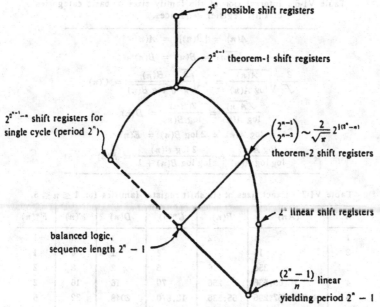

Fig. VI-8. Family sizes of shift register sequences (partially ordered by inclusion).

$$\binom{2^{n-1}}{2^{n-2}} \sim \frac{2}{\sqrt{\pi}} 2^{2^{n-1}-n/2},$$

which is only slightly larger than $2^{2^{n-1}-n}$. Among these are 2^n linear shift registers, of which $\phi(2^n - 1)/n$ yield two-cycle vector diagrams, the longer cycle containing $2^n - 1$ of the possible 2^n terms.

For purposes of comparison, it is convenient to give names to the six order functions which arise. Let

$$A(n) = 2^{2^n} \qquad\qquad D(n) = 2^{2^{n-1}-n}$$
$$B(n) = 2^{2^{n-1}} \qquad\qquad E(n) = 2^n$$
$$C(n) = \frac{2}{\sqrt{\pi}} 2^{(2^{n-n})/2} \qquad F(n) = \frac{2^n}{n}.$$

Table VI-6 expresses these functions in terms of $A(n)$ and $B(n)$. All logarithms are to the base 2. For all n, $A(n) > B(n) > C(n) > D(n) > E(n) > F(n)$. Many other relationships, of the type $A(n-1) = B(n)$ or $D(n+1) > B(n)$, can easily be derived.

The size of the shift register family having balanced logics and yielding sequences of length $2^n - 1$ is not known exactly, though it is

Table VI-6. Interrelation of the family sizes of basic categories of shift register sequences.

$$A(n) = [B(n)]^2 = A(n)$$

$$\sqrt{A(n)} = B(n) = B(n)$$

$$\frac{2}{\sqrt{\pi}}\sqrt{\frac{A(n)}{\log A(n)}} = 4\sqrt{\frac{2}{\pi}}\frac{B(n)}{\sqrt{\log B(n)}} = C(n)$$

$$\frac{\sqrt{A(n)}}{\log A(n)} = \frac{2(Bn)}{\log B(n)} = D(n)$$

$$\log A(n) = 2\log B(n) = E(n)$$

$$\frac{\log A(n)}{\log\log A(n)} = \frac{2\log B(n)}{\log\log B(n)+1} = F(n)$$

Table VI-7. Exact sizes of six shift register families for $1 \le n \le 5$.

n	(An)	$B(n)$	$C^*(n)$	$D(n)$	$E(n)$	$F^*(n)$
1	4	2	0	1	2	1
2	16	4	2	1	4	1
3	256	16	6	2	8	2
4	65,536	256	70	16	16	2
5	4,294,967,296	65,536	12,870	2048	32	6

believed to have the order $2^{2^{n-1}-(2n/2)}$. A comparison of the exact sizes of the other six families appears in Table VI-7 for $n \le 5$. Here, the approximate functions $C(n)$ and $F(n)$ of Table VI-6 have been replaced by the exact functions $C^*(n) = \binom{2^{n-1}}{2^{n-2}}$ and $F^*(n) = \phi(2^n - 1)/n$.

Short cycles. It is not difficult to obtain criteria for the presence or absence of various short cycles in the vector diagrams of shift registers. There are two possible cycles of length 1 (i.e., *all* 0's and *all* 1's), only one cycle of length 2 (i.e., 101010...), two cycles of length 3, and, in general, $(1/n)\sum_{d|n}\mu(d)2^{n/d}$ cycles of length n.

Theorem 11. The cycle *all* 0's occurs if and only if $F_1(0,0, \dots,0) = 0$, and *all* 1's is a cycle if and only if $F_1(1,1, \dots,1) = 0$. (These conditions are independent of the tap positions for the feedback logic.)

Theorem 12. The cycle 101010... occurs if $F_1(x_1, x_2, x_3, \dots) = F_1(x_1', x_2', x_3', \dots) \equiv n \pmod 2$, where n is the shift register degree, and the prime denotes complementation.[1]

[1] The condition given in Theorem 12 is sufficient, but not *necessary*, for the cycle (10) to occur. This question is more completely treated in Chapter VII.

These theorems follow directly from the truth-table ideas of Section 1.2. Progressively more complex criteria can be given for cycles of length 3, 4, etc.

Let $F_1 = x_1 \oplus x_2 \oplus \cdots \oplus x_k$ be a linear logic with k taps. Since $F_1(0, 0, \ldots, 0) = 0$, *all* 0's occurs as a cycle of length 1 for all linear logics. However, $F_1(1, 1, \ldots, 1) = 0$ if and only if k is even, so that *all* 1's is a cycle only for, those linear logics having an even number of taps (not counting the end tap w). Such logics always lead to highly degenerate vector diagrams. Finally, $F_1(x_1, x_2, \ldots, x_k) - F_1(x_1', x_2', \ldots, x_k') \equiv k \pmod 2$, so that $101010\ldots$ is a cycle provided that (1) k is even, and (2) the sum of the tap numbers has the same parity as n.

Suppose that the feedback logic $w \oplus y \oplus z \oplus xy \oplus xz \oplus yz$ is used. Here, $F_1 = y \oplus z \oplus xy \oplus xz \oplus yz$. Since $F_1(0, 0, 0) = 0$, and $F_1(1, 1, 1) = 1$, *all* 0's is always a short cycle, whereas *all* 1's never is. Moreover, $F_1(x', y', z') - F_1(x, y, z) = (y' - y) \oplus (z' - z) \oplus (x'y' - xy) \oplus (x'z' - xz) \oplus (y'z' - yz) = 1$, rather than 0, so that $101010\ldots$ is never a cycle for this logic.

Alternatively, suppose that the feedback logic $1 \oplus w \oplus y \oplus z \oplus xy \oplus yz$ is used. Here, $F_1(x, y, z) = 1 \oplus y \oplus z \oplus xy \oplus yz$, so that $F(0, 0, 0) = 1$, and $F(1, 1, 1) = 1$. Thus, no cycles of length 1 occur. However, $F_1(x', y', z') - F_1(x, y, z) = (y' - y) \oplus (z' - z) \oplus (x'y' - xy) \oplus (y'z' - yz) = x \oplus z$. Thus, a necessary condition for the occurrence of the cycle $101010\ldots$ is $x \equiv z \pmod 2$. The other condition is $1 \oplus y \oplus z \oplus xy \oplus yz \equiv n \pmod 2$, which reduces to $x \oplus y \equiv 1 \oplus n \pmod 2$. Therefore, the two congruences $x \equiv z$, $x \oplus y \equiv 1 \oplus n \pmod 2$ are necessary and sufficient for the existence of a two-cycle for the logic $1 \oplus w \oplus y \oplus z \oplus xy \oplus yz$.

Cycles of length n are specified by suitable congruences modulo n. The linear case is particularly simple.

A simplified statistical model.[1] It is interesting to inquire whether any simple statistical model can be used to approximate the situation of selecting a feedback logic with a random set of taps and recording the resulting cycle length. Such a model is described in the following paragraphs. The experimental results, presented in Section 3 of this chapter, provide ample justification for reliance on this particular model.

As observed previously, a shift register is merely a device for assigning successors to vectors in binary n-space Ω_n. Strictly speaking, it is demonstrably false to assert that selecting a shift register at random is equivalent to assigning successors to vectors at random. In fact, the successor V_2 of any vector V_1 has $n - 1$ components in common

---

[1] The "simplified statistical model" introduced briefly in Section 2.1 of this chapter is developed in considerable detail in Chapter VII.

with V_1, except for a shift in position. However, it is interesting to explore the consequences of accepting this random-successor postulate as a crude statistical approximation of the truth.

If attention is restricted to shift registers satisfying the conditions of Theorem 1 (cycles without branches), the corresponding simplified statistical model is as follows: Let V_0 be the initial state of the shift register. Let all 2^n vectors be placed in a hat. These are sampled without replacement, one at a time, until V_0 is drawn. This completes the cycle containing V_0.

Under these assumptions, it is not difficult to deduce a statistical distribution for the lengths of the cycles containing the given vector V_0 as the feedback logic of the shift register is varied "randomly."

> *Theorem 13.* The distribution of cycle lengths, for fixed initial vector and variable feedback logic, is a flat distribution for all lengths between 1 and 2^n.

Proof. The chance of picking the vector V_0 on the first try from the set of all 2^n vectors is clearly $1/2^n$.

Suppose that the probability of picking V_0 on the jth try is known to equal $1/2^n$ for $1 \leq j \leq k$. Then, the probability that the cycle has not yet ended by the $(k + 1)$st trial is $1 - (k/2^n)$. The probability of picking V_0 on the $(k + 1)$st trial, assuming that the cycle has continued this far, is $1/(2^n - k)$, since only $2^n - k$ vectors remain in the sample set. Thus, the probability that the cycle beginning with V_0 has length exactly $k + 1$ is the product $[1 - (k/2^n)][1/(2^n - k)] = 1/2^n$. This proves by complete induction that every cycle length from 1 to 2^n has probability $1/2^n$, so that the expected distribution of cycle lengths is a flat distribution.

The experimental evidence compiled thus far strongly supports the contention that the cycle-length distribution is flat. A detailed description of the experimental work appears in Section 3. It is interesting to note, in Section 4.1, not only that the 180 samples are rather flatly distributed between 0 and 2^{11}, but also that the number of coincidences (two cycles of the same length) agrees perfectly with the expected number of boxes containing 2 objects, when 180 objects are distributed randomly among 2048 boxes: that is, $2048 \binom{180}{2} [(2047^{178})/(2048^{180})] \simeq 7$.

2.2. *Maximum-length sequences*

The discussion which follows deals with sequences of maximum length and presents a method for obtaining all such sequences.

Good's diagram. To facilitate the study of the situation wherein all 2^n binary n-vectors lie on a single shift register cycle, Good [17] introduced an important vector diagram. All 2^n binary n-vectors are portrayed as nodes of an oriented graph, and arrows lead from each vector to all its possible shift register successors. Since the vector $(a_1, a_2, \ldots, a_n)$ has two possible successors, $(a_2, \ldots, a_n, 0)$ and $(a_2, \ldots, a_n, 1)$, as well as two possible predecessors, $(0, a_1, \ldots, a_{n-1})$ and $(1, a_1, \ldots, a_{n-1})$, there are two arrows entering and two arrows leaving every node of Good's diagram. The actual diagrams for $n = 1, 2, 3, 4$ appear in Figure VI-9.[*]

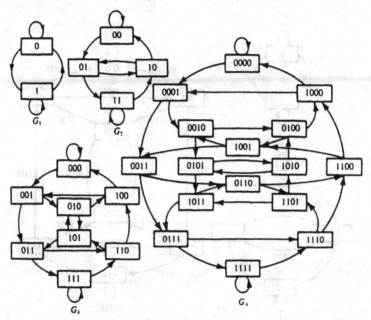

Fig. VI-9. Good's diagrams for $n = 1$, $n = 2$, $n = 3$, $n = 4$.

By the nature of the construction, if two vectors have one successor in common, they have both successors in common. The first three diagrams are seen in Figure VI-9 to be planar. However, for $n = 4$, the diagram is easily shown to be nonplanar.

A shift register sequence of length 2^n corresponds to a closed path through Good's diagram G_n which traverses each node exactly once. It also corresponds to a complete path in G_{n-1} (that is, a closed path

---

[*] Called "de Bruijn diagrams" in Chapter II. Priority between Refs. [3] and [17] is hard to establish, as both appeared in 1946.

in G_{n-1} which traverses each edge exactly once and in the positive direction), because each edge of G_{n-1} may be identified with the n-vector whose first $n-1$ components are the same as the terminal node of that edge. This is illustrated in Figure VI-10. In [17], Good proved the following Theorem:

> *Theorem 14.* Let G be a connected oriented graph such that, at each node, the number of edges oriented toward the node equals the number of edges oriented away from the node. Then G possesses a complete path.

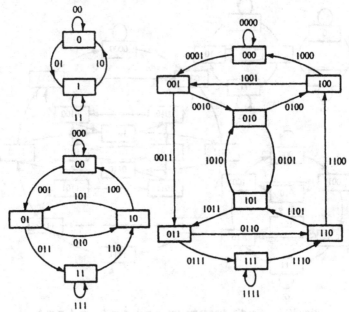

Fig. VI-10.   The edges of Good's diagrams considered as $(n+1)$-vectors.

Applying this theorem to G_n, Good shows the existence of shift register sequences of period 2^n, admitting, "I know no simple formula for the number of solutions." An alternative existence proof was given by Rees [30], who effectively constructed a linear shift register sequence of length $2^n - 1$ and inserted an extra 0 in the longest run of 0's to obtain length 2^n.

More significant use of Good's diagram was made by de Bruijn [3], who apparently discovered the diagram independently. In [3], de Bruijn succeeded in proving the following theorem.

> *Theorem 15.* The exact number of complete paths in G_{n-1} is $2^{2^{n-1}-n}$ (i.e., the number of sequences of length 2^n obtainable from n-stage shift registers).

De Bruijn's proof consists of reducing the problem to the problem of showing that the number of complete paths $|G_n|$ in G_n satisfies the relation $|G_{n+1}| = 2^{2^{n-1}-1}|G_n|$. He then proves this in the more general framework of oriented graphs and a *doubling* operation. His proof is essentially a counting argument and gives no effective routine for constructing complete paths. The section on preference functions, below, contains a more constructive proof.

Henceforth, maximum-length shift register sequences are referred to as de Bruijn sequences.

Preference functions. The preference-function technique discussed here produces sequences of degree n from a different approach. The importance of this technique lies in the fact that, although little is known about the relation between the de Bruijn sequences and the feedback formulas, fairly simple methods of constructing preference functions yield all de Bruijn sequences. The drawback to this technique is the large amount of storage needed in any mechanization of the generation of sequences from preference tables. It is assumed that the digits in a sequence are taken from the t-alphabet $0, 1, \ldots, t-1$ and are later specialized to the binary alphabet.

Definition 1. The statement that P is a preference function means that P is a t-dimensional vector-valued function of $n-1$ variables such that, for each choice of $a_1, a_2, \ldots, a_{n-1}$ from the t-alphabet, $[P_1(a_1, \ldots, a_{n-1}), \ldots, P_t(a_1, \ldots, a_{n-1})]$ is a rearrangement of $0, \ldots, t-1$.

Examples of preference functions are:

1. The constant function $P(x_1, \ldots, x_{n-1}) = (0, 1, \ldots, t-1)$.
2. The cyclic permutation $P(x_1, \ldots, x_{n-1}) = (x_1 + 1, x_1 + 2, \ldots, x_1 + t) \bmod t$.

Definition 2. For any n-digit word $(I_1, \ldots, I_n)$ and preference function P, the following inductive definition determines a unique finite sequence $\{a_i\}$:

1. $a_1 = I_1, \ldots, a_n = I_n$.
2. If $a_{N+1}, \ldots, a_{N+n-1}$ have been defined, then $a_{N+n} = P_i(a_{N+1}, \ldots, a_{N+n-1})$, where i is the smallest integer such that the word $[a_{N+1}, \ldots, a_{N+n-1}, P_i(a_{N+1}, \ldots, a_{N+n-1})]$ has not previously occurred as a segment of the sequence (provided that there are such i).
3. Let $L\{a_i\}$ be the first value of N such that no i can be found to satisfy item 2. Then, a_{L+n-1} is the last digit of the sequence and

$L\{a_i\}$ is called the cycle period.

> *Lemma 1.* The sequence defined above is recursive of degree n. That is, each n-digit word occurs, at most, once.

Proof. The Lemma is an immediate consequence of Definition 2 (2).

> *Lemma 2.* The sequence described in Definition 2 is cyclic of period $L\{a_i\}$, in the sense that $(a_1, \ldots, a_{n-1}) = (a_{L+1}, \ldots, a_{L+n-1})$. The significance of this Lemma is that, if $S_N = (a_N, \ldots, a_{N+n-1})$ are the successive states of a shift register without branch points (see Section 2.1), then S_1 follows S_L.

Proof. Since $(a_{L+1}, \ldots, a_{L+n-1})$ is the terminal $(n-1)$-digit word, every n-digit word of the form $(a_{L+1}, \ldots, a_{L+n-1}, x)$ appears as a segment of $\{a_i\}$. Therefore, $(a_{L+1}, \ldots, a_{L+n-1})$ occurs $t+1$ times, altogether. If each of these $t+1$ occurrences were preceded by a digit, then Lemma 1 would be false. Thus, $(a_{L+1}, \ldots, a_{L+n-1})$ must occur once without a predecessor, and this can only be $(a_1, \ldots, a_{n-1})$.

Hereafter, no distinction is made between the finite sequence and its periodic extension of period $L\{a_i\}$.

> *Lemma 3.* For any cyclic recursive sequence $\{a_i\}$ of degree n and for any n-digit word in it, there exists a preference function which generates the sequence, using the given word as the initial word.

Proof. For any $(n-1)$-digit word $(x_1, \ldots, x_{n-1})$, let $P_i(x_1, \ldots, x_{n-1})$ be the successor of the ith occurrence of $(x_1, \ldots, x_{n-1})$ in $\{a_i\}$, starting from the initial word. If there are less than i occurrences, fill in the remaining components of $P(x_1, \ldots, x_{n-1})$ arbitrarily, subject to the restriction that the components are all distinct.

For $(x_1, \ldots, x_{n-1}) \neq (I_1, \ldots, I_{n-1})$, the ith occurrence of $(x_1, \ldots, x_{n-1})$ in a de Bruijn sequence is followed by $P_i(x_1, \ldots, x_{n-1})$. The first occurrence of $(I_1, \ldots, I_{n-1})$ is followed by I_n and is followed thereafter by the other digits in their relative order in $P(I_1, \ldots, I_{n-1})$. Since every $(n-1)$-digit word occurs t times, it is easily seen that the following Lemma is valid.

> *Lemma 4.* For a given initial word, there is a one-to-one correspondence between de Bruijn sequences of degree n and their preference functions which satisfy $P_1(I_1, \ldots, I_{n-1}) = I_n$.

It is now possible to describe some methods of constructing preference functions which yield de Bruijn sequences. The first example is constructed from the initial word $(I_1, \ldots, I_n)$.

Definition 3. The statement that $(x_1, \ldots, x_{n-1})$ has an r-overlap with $(I_1, \ldots, I_n)$ means that $0 \leq r \leq n - 1$ and $(x_{n-r}, \ldots, x_{n-1}) = (I_1, \ldots, I_r)$. Notice that every $(n - 1)$-digit word has a zero overlap with $(I_1, \ldots, I_n)$.

> *Theorem 16.* Theorem 16 is herein designated as the *key-word theorem* (KWT).[4] For any initial word $(I_1, \ldots, I_{n-1})$, if P satisfies $P_t(x_1, \ldots, x_{n-1}) = I_{r+1}$, where r is the largest integer such that $(x_1, \ldots, x_{n-1})$ has an r-overlap with $(I_1, \ldots, I_n)$, then the sequence generated by $(I_1, \ldots, I_n)$ and P has length t^n.

Proof. It is now shown by induction on k that $(l_k, \ldots, l_1, I_1, \ldots, I_{n-k})$ is in the sequence for all choices of $l_1, \ldots, l_k$. For $k = n$, this says that every n-digit word is in the sequence, and, since the sequence is recursive of degree n, there must be $t^n + n - 1$ digits; this means that the cycle period is t^n. There are two steps in the inductive proof:

1. The case $k = 1$. By Lemma 2, the terminal $(n - 1)$-digit word of the sequence is $(I_1, \ldots, I_{n-1})$. Since this is the terminal $(n - 1)$-digit word, it must have occurred $t + 1$ times. One occurrence is $(a_1, \ldots, a_{n-1})$, and the remaining t occurrences have a predecessor $(l_1, I_1, \ldots, I_{n-1})$. Since these are distinct words, every choice of l_1 occurs.

2. The case $n \geq k \geq 2$. By the inductive hypothesis, $(l_{k-1}, \ldots, l_1, I_1, \ldots, I_{n-k+1})$ is in the sequence for every choice of $(l_1, \ldots, l_{k-1})$. Now, if $(l_{k-1}, \ldots, l_1, I_1, \ldots, I_{n-k})$ has an overlap with the initial word greater than $n - k$ for some choice of $l_1, \ldots, l_k$, then, by inductive hypothesis, $(l_k, \ldots, l_1, I_1, \ldots, I_{n-k})$ is in the sequence for every choice of l_k. If $n - k$ is the greatest overlap with the initial word and $k \geq 2$, then it follows that $(l_{k-1}, \ldots, l_1, I_1, \ldots, I_{n-k}) \neq (I_1, \ldots, I_{n-1})$; the ith occurrence is followed by $P_i(l_{k-1}, \ldots, l_1, I_1, \ldots, I_{n-k})$; and $P_i(l_{k-1}, \ldots, l_1, I_1, \ldots, I_{n-k}) = I_{n-k+1}$. Since, by the inductive hypothesis, there is an occurrence followed by I_{n-k+1}, there must be t occurrences. Each of these will have a distinct predecessor, so that $(l_k, l_{k-1}, \ldots, l_1, I_1, \ldots, I_{n-k})$ occurs for all choices of l_k.

As an application of Theorem 16, the following example may be cited: Let $t = 3$, $n = 3$, and the key word be (012). The preference function given by Table VI-8 satisfies the conditions of KWT and yields a de Bruijn sequence of cycle length 27 (sequence length 29).

The key-word method gives $t^n[(t - 1)!]t^{n-1}$ maximum-length finite sequences and possibly less than this number of periodic sequences, if sequences are equated which differ only by some shift. The key-

---

[4] The key-word theorem is a direct generalization of a theorem developed by Ford [6]. Ford's initial word was $I_g = t - 1$ for $g = 1, \ldots, n$ and $P_i(x_1, \ldots, x_{n-1}) = i - 1$ for $i = 1, \ldots, t$.

Table VI-8. Preference table and resulting sequence (Theorem 16).

x_1	x_2	P_1	P_2	P_3
0	0	2	0	1
0	1	1	0	2
0	2	1	2	0
1	0	0	2	1
1	1	1	2	0
1	2	2	1	0
2	0	0	2	1
2	1	2	1	0
2	2	2	1	0

$\{a_i\}$ = 0122212111200210001102202010 1

sequence method discussed below gives a much larger class. Although the following theorem can be extended to the t-alphabet, the proof is somewhat simpler using a 2-alphabet. The generalization to the t-alphabet will not be given here.

Theorem 17. This theorem is designated here as the *key-sequence theorem* (KST). The following hypotheses are adopted:

1. Let $(I_1, \ldots, I_n)$ be an arbitrary initial word.

2. Let $P(x_1, \ldots, x_{n-2}) = (P_1, P_2)$ be the preference function for the binary de Bruijn sequence of degree $n - 1$, $\{b_i\}$, and initial word $(I_1, \ldots, I_{n-1})$, such that $P_1(I_1, \ldots, I_{n-2}) = 1 \oplus I_{n-1} \bmod 2$.

3. Let $x_1 \oplus F(x_2, \ldots, x_{n-1})$ be the feedback formula for $\{b_i\}$.

4. Let $P^*(x_1, \ldots, x_{n-1}) = (P_1^*, P_2^*)$ which satisfies $P_2^* = 1 \oplus P_1^*$ and $0 = [P_1^*(x_1, \ldots, x_{n-1}) \oplus 1 \oplus x_1 \oplus F(x_2, \ldots, x_{n-1})] \times [x_1 \oplus F(x_2, \ldots, x_{n-1}) \oplus P_1(x_2, \ldots, x_{n-1})]$.

It follows from these hypotheses that the sequence $\{a_i\}$, generated by $(I_1, \ldots, I_n)$ and P^*, is de Bruijn of degree n.

Proof. The proof consists of showing by induction that $(b_{N+1}, \ldots, b_{N+n-1})$ is contained in a cycle of $\{a_i\}$ twice for $1 \le N \le 2^{n-1}$. Since $\{a_i\}$ is recursive of degree n, and since every $(n - 1)$-digit word is contained in $\{b_i\}$, the maximality of $\{a_i\}$ follows. There are two steps in the inductive proof:

1. The case $N = 2^{n-1}$. The proof of Lemma 2 shows that $(I_1, \ldots, I_{n-1})$ occurs twice in a cycle of $\{a_i\}$. Since $(b_{N+1}, \ldots, b_{N+n-1}) = (I_1, \ldots, I_{n-1})$, the inductive hypothesis is true for $N = 2^{n-1}$.

2. The case $2^{n-1} > N \ge 1$. Assume the inductive hypothesis for all i such that $2^{n-1} \ge i \ge N + 1$. It is desired to prove this hypothesis for $i = N$. Since $\{b_i\}$ is de Bruijn, there exists M such that $(b_{M+1}, \ldots,$

$b_{M+n-1}) = (b_{N+2}, \ldots, b_{N+n-1}, 1 \oplus b_{N+n})$. If $2^{n-1} \geq M > N + 1$, then, by the inductive hypothesis, both $(b_{N+2}, \ldots, b_{N+n})$ and $(b_{N+2}, \ldots, b_{N+n-1}, 1 \oplus b_{N+n})$ have occurred twice in a cycle of $\{a_i\}$. Therefore, $(b_{N+2}, \ldots, b_{N+n-1})$ has occurred four times, and this implies that $(b_{N+1}, \ldots, b_{N+n-1})$ occurs twice.

On the other hand, suppose that $1 \leq M \leq N + 1$. If $N + 1 = 2^{n-1}$, by hypothesis 3, then $P_1(b_{N+2}, \ldots, b_{N+n-1}) = 1 \oplus I_{n-1} = 1 \oplus b_{N+n}$. If $2 \leq N + 1 < 2^{n-1}$, then $P_1(b_{N+2}, \ldots, b_{N+n-1}) = b_{M+n-1} = 1 \oplus b_{N+n}$. Thus, in either event, $P_1(b_{N+2}, \ldots, b_{N+n-1}) = 1 \oplus b_{N+n} = 1 \oplus b_{N+1} \oplus F(b_{N+2}, \ldots, b_{N+n-1})$. By hypothesis 4, this implies that

$$P_1^*(b_{N+1}, \ldots, b_{N+n-1}) = 1 \oplus b_{N+1} \oplus F(b_{N+2}, \ldots, b_{N+n-1}) = 1 \oplus b_{N+2}.$$

Therefore, the first occurrence of $(b_{N+1}, \ldots, b_{N+n-1})$ is followed by $1 \oplus b_{N+n}$. However, the inductive hypothesis asserts that $(b_{N+2}, \ldots, b_{N+n})$ occurs twice in a cycle of $\{a_i\}$, which implies that $(b_{N+1}, \ldots, b_{N+n-1}, b_{N+n})$ occurs in $\{a_i\}$. Thus, $(b_{N+1}, \ldots, b_{N+n-1})$ must occur twice in a cycle of $\{a_i\}$. This completes the induction.

The following example is given as an application of Theorem 17: Let $(b_1, \ldots, b_8) = 11100010$ and $(I_1, I_2, I_3, I_4) = (1111)$. The preference functions P_1^{*i} and the corresponding de Bruijn sequences are listed in Table VI-9.

Table VI-9. Preference tables and resulting sequences (Theorem 17).

x_1	x_2	x_3	P_1^*	P_1^{*1}	P_1^{*2}	P_1^{*3}	P_1^{*4}	P_1^{*5}	P_1^{*6}	P_1^{*7}	P_1^{*8}
0	0	0	0	0	0	0	0	0	0	0	0
0	0	1	x	0	1	0	1	0	1	0	1
0	1	0	0	0	0	0	0	0	0	0	0
0	1	1	0	0	0	0	0	0	0	0	0
1	0	0	x	0	0	1	1	0	0	1	1
1	0	1	0	0	0	0	0	0	0	0	0
1	1	0	x	0	0	0	0	1	1	1	1
1	1	1	1	1	1	1	1	1	1	1	1

Sequences

1. 1111000010011010111
2. 1111000011010010111
3. 1111001000011010111
4. 1111001101000010111

5. 1111010000101100111
6. 1111010000110010111
7. 1111010010110000111
8. 1111010011000010111

Corollary to Theorem 17. If N_{n-1} is the number of de Bruijn sequences of degree $n - 1$, and if $(I_1, \ldots, I_n)$ is an initial word, then the key-sequence method yields $2^{2^{n-2}-1} N_{n-1}$

distinct de Bruijn sequences of degree n.

Proof. $P_1^*(x_1, \ldots, x_{n-1})$ is unrestricted under the condition that $0 = x_1 \oplus F(x_2, \ldots, x_{n-1}) \oplus P_1(x_2, \ldots, x_{n-1})$. Thus, there are $2^{2^{n-2}}$ preference functions satisfying the conditions of KST for each de Bruijn sequence of degree $n - 1$. Now, the fact that $0 = I_1 \oplus F(I_2, \ldots, I_{n-1}) \oplus P_1(I_2, \ldots I_{n-1})$ follows from the properties of preference functions if $(I_1, \ldots, I_{n-1})$ is not the *all* 0's or the *all* 1's word, and, otherwise, follows from hypothesis 2 of KST. Therefore, there are $2^{2^{n-2}}/2$ preference functions which satisfy the condition of Lemma 4, so that there are $2^{2^{n-2}}/2$ de Bruijn sequences of degree n associated with each de Bruijn sequence of degree $n - 1$.

Next, suppose that $\{a_i\}$ can be obtained from KST, from both $\{b_i\}$ and $\{c_i\}$ (distinct de Bruijn sequences of degree $n - 1$). Let $b_i = c_i = I_i$ for $i = 1, \ldots, n - 1$, and let P^1, P^2 be their preference functions which satisfy hypothesis 2 of KST. Finally, let N be the smallest integer $\geq -n + 1$, such that $b_{N+n-1} \neq c_{N+n-1}$. Then, $N > 0$ and $P_1^1(b_{N+1}, \ldots, b_{N+n-2}) = b_{N+n-1} = 1 \oplus c_{N+n-1} = 1 \oplus P_1^2(b_{N+1}, \ldots, b_{N+n-2})$. Let $x_1 = 1 \oplus b_N = 1 \oplus c_N$ and $x_{i+1} = b_{N+i} = c_{N+i}$ for $i = 1, \ldots, n - 2$. Then, $x_1 \oplus F^k(x_2, \ldots, x_{n-1}) \oplus P_1^k(x_2, \ldots, x_{n-1}) = 1$ for $k = 1, 2$. Therefore, hypothesis 4 of KST implies that $P_1^*(x_1, \ldots, x_{n-1}) = b_{N+n-1}$ and $P_1^*(x_1, \ldots, x_{n-1}) = c_{N+n-1}$, which is a contradiction.

Theorem 18 (*de Bruijn's theorem*). Let N_{n-1} be the number of de Bruijn sequences of degree $n - 1$. Then, there are $2^{2^{n-2}-1}N_{n-1}$ de Bruijn sequences of degree n.

Proof. In view of the corollary to KST, it must be shown that, for each de Bruijn sequence of degree n, there exists one sequence of degree $n - 1$ which satisfies the conditions of KST. Let $(I_1, \ldots, I_n)$ be the initial word and $\{a_i\}$ the de Bruijn sequence of degree n starting at $a_1 = I_1, \ldots, a_n = I_n$. The sequence $\{b_i\}$ is defined as follows:

1. Let $b_1 = I_1, \ldots, b_{n-1} = I_{n-1}$.
2. Suppose that $N \geq 1$ and $b_1, \ldots, b_{N+n-2}$ have been defined. Then:
   a. If $(b_{N+1}, \ldots, b_{N+n-2}) = (b_{x+1}, \ldots, b_{x+n-2})$ has two solutions with $x < N$, the sequence terminates.
   b. If the equation has one solution with $x < N$, define $b_{N+n-1} = b_{x+n-1} \oplus 1$.
   c. If there are no solutions, then consider the two solutions x_1, x_2 to the equation $(1 \oplus b_N, b_{N+1}, \ldots, b_{N+n-2}) = (a_x, a_{x+1}, \ldots, a_{x+n-2})$. Let x_1 be the smaller of x_1, x_2. Then define $b_{N+n-1} = a_{x_1+n-1}$. This definition is analogous to a construction from a preference function, and Lemmas 1 and 2 are valid for $\{b_i\}$.

Next, it is asserted that $\{b_i\}$ is de Bruijn. Suppose that there exists a word $(x_1, \ldots, x_{n-1})$ which is not in the sequence $\{b_i\}$. Since every $(n-1)$-digit word occurs twice in $\{a_i\}$, one can choose N to be the largest integer such that $(a_N, \ldots, a_{N+n-2})$ is not in $\{b_i\}$, and $N \le 2^{n-1}$. The other solution to the equation $(a_z, \ldots, a_{z+n-2}) = (a_N, \ldots, a_{N+n-2})$ has the property that $x < N$. Because of the choice of N, if the $(n-2)$-digit word $(a_{N+1}, \ldots, a_{N+n-2})$ occurs in $\{b_i\}$, it must be preceded by $1 \oplus a_N$. Then, by definition 2(c), above, it must be followed by a_{z+n-1}, which is $a_{N-n-1} \oplus 1$. Therefore, $(a_{N+1}, \ldots, a_{N+n-2}, a_{N+n-2}, a_{N+n-1})$ cannot occur in $\{b_i\}$. If $N = 2^{n-1}$, this leads to the contradiction that $(I_1, \ldots, I_{n-1})$ is not in $\{b_i\}$. If $N < 2^{n-1}$, this contradicts the maximality of N.

Finally, it is asserted that hypothesis 4 of KST is satisfied. Consider the words $(x_1, \ldots, x_{n-1}) = (b_{N_1}, b_{N_1+1}, \ldots, b_{N_1+n-2})$, and $(1 \oplus x_1, x_2, \ldots, x_{n-1}) = (b_{N_2}, b_{N_2+1}, \ldots, b_{N_2+n-2})$. If $1 < N_1 < N_2 \le 2^{n-1}$, then $x_1 \oplus F(x_2, \ldots, x_{n-1}) = P_1(x_2, \ldots; x_{n-1})$, and hypothesis 4 is satisfied. If $1 \le N_2 < N_1$, then, by the definition of $\{b_i\}$, $b_{N_2} \oplus F(b_{N_2+1}, \ldots, b_{N_2+n-2}) = b_{N_2+n-1} = a_{M+n-1}$, where M is the first occurrence of $(1 \oplus b_{N_2}, b_{N_2+1}, \ldots, b_{N_2+n-2})$ in $\{a_i\}$. Thus, $P_1^*(1 \oplus b_{N_2}, b_{N_2+1}, \ldots, b_{N_2+n-2}) = b_{N_2} \oplus F(b_{N_2+1}, \ldots, b_{N_2-n-2})$, which is $P_1^*(x_1, \ldots, x_{n-1}) = 1 \oplus x_1 \oplus F(x_2, \ldots, x_{n-1})$, and hypothesis 4 is satisfied.

The key-sequence method can be extended to yield de Bruijn sequences of degree n from de Bruijn sequences of degree $k < n-1$, provided that $(I_1, \ldots, I_n)$ is a word in the lower-order sequence. The set of all de Bruijn sequences of degree k yields 2 to the exponent $(2^{n-2} - 2^{n-k-1} + 2^{k-1} - k)$ sequences of degree n.

Subgroups of the feedback groups. It would be convenient to have some simple methods for obtaining large classes of feedback formulas which yield de Bruijn sequences. The discussion which follows deals with an attempt to develop one such method.

There is a simple composition operation under which the feedback functions satisfying the conditions of Theorem 1 form a group. Remembering that such feedbacks may be written in the form $x_1 \oplus F(x_2, \ldots, x_n) = x_{n+1}$, and calling F the reduced function, one obtains the following Theorems.

Theorem 19. The $2^{2^{n-1}}$ feedback functions of degree n satisfying the branchless-cycles condition (Theorem 1) form a commutative group under the operation of termwise addition modulo 2 of the reduced functions. In this group, the cycling shift register acts as the identity, and all other group elements have order 2. This group may conveniently be designated as the feedback group.

Proof. Routine verification of the group postulates constitutes proof of Theorem 19.

Theorem 20. Any of the 2^{n-1} constraints $F(a_2, \ldots, a_n) = 0$ defines a subgroup of index 2 in the feedback group, as does the constraint that the cycle number be even. The simultaneous imposition of k of these constraints, for $1 \leq k \leq 2^{n-1}$, defines a subgroup of index 2^k.

Proof. If two functions F_1 and F_2 both vanish for the vector $V = (v_2, \ldots, v_n)$, so also does their sum $F_1 \oplus F_2$. Thus, the condition of vanishing at V defines a subgroup. Since one-half of the $2^{2^{n-1}}$ functions in question vanish at V, the subgroup has index 2.

By Theorem 5, an even cycle number is equivalent to an even number of 1's in the truth table of F: i.e., to the truth table summing to 0 modulo 2. This property is clearly preserved in the addition $F_1 \oplus F_2$. One-half of the functions have truth tables which sum to zero modulo 2, whence this subgroup also has index 2.

The property of F vanishing at V_1 is statistically independent of F vanishing at V_2 or of the output table of F summing to zero. Thus, a simultaneous imposition of k such linear constraints, for $1 \leq k \leq 2^{n-1}$, defines a subgroup of index 2^k.

The imposition of n constraints on F yields a subgroup of index 2^n and, hence, of order $2^{2^{n-1}}/2^n = 2^{2^{n-1}-n}$, which is de Bruijn's number. All the cosets of the subgroup in question also have $2^{2^{n-1}-n}$ elements.

For $n = 1, 2, 3$, the de Bruijn sequences are characterized by the following conditions:

1. $F(0, \ldots, 0) = 1$
2. $F(1, \ldots, 1) = 1$
3. $\sum_{V} F(V) \equiv 1 \,(\text{mod } 2)$

Therefore, for $n = 1, 2, 3$, the reduced functions for de Bruijn sequences form a coset of the subgroup defined by setting these three expressions equal to zero. For $n = 1, 2$, the subgroup is of order 1, and, for $n = 3$, the subgroup is of order 2.

It is natural to ask whether, in general, there is a subgroup of which the de Bruijn feedbacks form a coset. The answer is that there is not, for $n \geq 4$. A more natural question is: for what subgroups do the de Bruijn feedbacks form a union of cosets, and what is the maximal subgroup having this property? It can be shown that the maximal subgroup will contain the others. Before these questions are answered, a notation is introduced: $\Phi_V(x_2, \ldots, x_n)$ is the reduced function which equals 1 if and only if $(x_2, \ldots, x_n) = V$.

Theorem 21. If $n \geq 3$, the feedbacks $x_1 \oplus \Phi_{(1,0,\ldots,0)}(x_2, \ldots, x_n) \oplus \Phi_{(0,\ldots,0,1)}(x_2, \ldots, x_n)$ and $x_1 \oplus \Phi_{(0,1\ldots11)}(x_2, \ldots, x_n) \oplus \Phi_{(1,\ldots,1,0)}(x_2, \ldots, x_n)$, regarded as group elements, leave the de Bruijn property invariant. Therefore, the de Bruijn feedbacks form a union of cosets of the subgroup generated by these two feedbacks.

Proof. The second statement follows from the first by elementary group theory. To prove the first statement, refer to Good's diagram G_{n-1}. A de Bruijn sequence of degree n is a complete path on this oriented graph. For each de Bruijn sequence, there are two paths away from the node $(0, \ldots, 0, 1)$ toward the node $(1, 0, \ldots, 0)$. These two paths are then connected by two paths; one of these goes directly from $(1, 0, \ldots, 0)$ to $(0, \ldots, 0, 1)$, and the other goes by way of $(0, \ldots, 0)$. There are two ways of connecting these paths to give de Bruijn sequences (provided that $n \geq 3$). If $x_1 \oplus F(x_2, \ldots, x_n)$ is the feedback of one sequence, then $x_1 \oplus F(x_2, \ldots, x_n) \oplus \Phi_{1,0,\ldots,0}(x_2, \ldots, x_n) \oplus \Phi_{(0,\ldots,0,1)}(x_2, \ldots, x_n)$ is the other.

This expression is obtained by applying the group operation to $x_1 \oplus F(x_2, \ldots, x_n)$ and $x_1 \oplus \Phi_{1,0,\ldots,0}(x_2, \ldots, x_n) \oplus \Phi_{(0,\ldots,0,1)}(x_2, \ldots, x_n)$. A similar argument holds for the other feedback.

Theorem 22. The subgroup generated by the two feedbacks listed in Theorem 21 is maximal.

Proof. Suppose that $x_1 \oplus f(x_2, \ldots, x_n)$ is in the maximal subgroup, but is not of the form $x_1 \oplus a[\Phi_{1,0,\ldots,0}(x_2, \ldots, x_n) \oplus \Phi_{0,\ldots,0,1}(x_2, \ldots, x_n)] \oplus b[\Phi_{0,1,\ldots,1}(x_2, \ldots, x_n) \oplus \Phi_{;,\ldots,1,0}(x_2, \ldots, x_n)]$ for any choice of the constants a and b. Then, one can choose a and b so that $f^1 = f \oplus a(\Phi_{1,0,\ldots,0} \oplus \Phi_{0,\ldots,0,1}) \oplus b(\Phi_{0,1,\ldots,1} \oplus \Phi_{1,\ldots,1,0})$ is zero for $(x_2, \ldots, x_n) = (1, 0, \ldots, 0)$, $(1, \ldots, 1, 0)$.

It will also be zero for $(1, \ldots, 1)$ and $(0, \ldots, 0)$. The feedback $x_1 \oplus f^1$ will be a member of the maximal subgroup other than the identity. It is now shown that $x_1 \oplus f^1(x_2, \ldots, x_n)$ is not in the maximal subgroup, and, thus, a contradiction is established.

Since $x_1 \oplus f^1(x_2, \ldots, x_n)$ is not the identity, there exists a vector $(a_2, \ldots, a_n)$ such that $f^1(a_2, \ldots, a_n) = 1$. Let

$$Q_k = (a_k, \ldots, a_n, a_2, \ldots, a_{k-1})$$
$$Q_k^* = (a_{k-1}, a_k, \ldots, a_n, a_2, \ldots, a_{k-1}).$$

$$(k = 2, \ldots, n)$$

Let i_k $(k = 1, \ldots, r)$ be those indices such that $f^1(Q_{i_k}) = 1$, and let j_k $(k = 1, \ldots, n - 1 - r)$ be indices such that $f^1(Q_{j_k}) = 0$. Since $f^1(Q_2) = 1$, then $r \geq 1$. Now, suppose that one can find a de Bruijn sequence with recursion $x_1 \oplus F(x_2, \ldots, x_n)$, so that $Q_{j_k}^*$ is followed by $Q_{j_{k+1}}^*$, for $k = 1, \ldots, n - 1 - r$, and $Q_{i_k}^*$ is not followed by $Q_{i_{k+1}}^*$, for $k = 1, \ldots, r$. It is easily seen that $x_{n+1} = x_1 \oplus F(x_2, \ldots, x_n) \oplus f^1(x_2, \ldots, x_n)$ has $(a_2, \ldots, a_n)$ as a cycle and, thus, is not the recursion of a de Bruijn sequence. Hence, $x_1 \oplus f^1$ cannot be in the maximal subgroup.

Thus, the contradiction is established if the existence of a de Bruijn

sequence with the above properties can be shown. As mentioned previously, a de Bruijn sequence of degree n can be represented as a complete walk on Good's diagram G_{n-1}. Certain nodes are now modified
to guarantee that the resulting path will have the desired properties.

The node corresponding to Q_{j_k} is replaced by two nodes, as shown
in Figure VI-11(a). Also, the node corresponding to Q_{i_k} is replaced by
two nodes, as shown in Figure VI-11(b). It is obvious that a complete
path on the resulting oriented graph will have the properties sought.
Good's theorem (Theorem 14) asserts that a complete path exists, provided that the oriented graph is connected, and that each node has as
many output as input edges. The unmodified nodes have 2 input and
2 output edges, whereas the modified nodes have 1 input and 1 output.
Thus, the proof will be complete if connectedness can be proved. This
is done in three steps. It is shown (1) that every modified node is
connected to an unmodified node; (2) that each unmodified node is connected to either $(0, \ldots, 0)$ or $(1, \ldots, 1)$; and (3) that $(0, \ldots, 0)$ is connected to $(1, \ldots, 1)$.

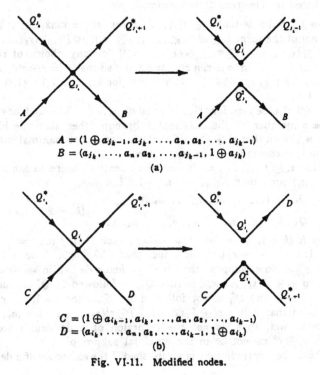

$$A = (1 \oplus a_{j_k-1}, a_{j_k}, \ldots, a_n, a_1, \ldots, a_{j_k-1})$$
$$B = (a_{j_k}, \ldots, a_n, a_1, \ldots, a_{j_k-1}, 1 \oplus a_{j_k})$$

(a)

$$C = (1 \oplus a_{i_k-1}, a_{i_k}, \ldots, a_n, a_1, \ldots, a_{i_k-1})$$
$$D = (a_{i_k}, \ldots, a_n, a_1, \ldots, a_{i_k-1}, 1 \oplus a_{i_k})$$

(b)

Fig. VI-11. Modified nodes.

Now, the edge $(1 \oplus a_{k-1}, a_k, \ldots, a_n, a_2, \ldots, a_{k-1})$ connects the node $(1 \oplus a_{k-1}, a_k, \ldots, a_n, a_2, \ldots, a_{k-2})$ to Q_k^2, as shown in Figure VI-11, and the former node cannot be a cyclic permutation of $(a_2, \ldots, a_n)$, since it does not have the same number of 1's among its components. If k is of the form i_m, then Q_k^1 is directly connected through $(a_k, \ldots, a_n, a_2, \ldots, a_{k+1})$ to an unmodified node, as shown in Figure VI-11(b). If k is of the form j_m, then Q_k^1 is connected to Q_{k+1}^1, as shown in Figure VI-11(a). This chain of connections can be extended at most $n - 1 - r$ edges $(< n - 1)$ before connecting to a $Q_{k'}^1$, where k' is of the form i_m; one more step then connects to an unmodified node.

Next, suppose that $(x_2, \ldots, x_n)$ is not a cyclic permutation of $(a_2, \ldots, a_n)$ and has a number of 1's among its components which is less than or equal to the number of 1's among $(a_2, \ldots, a_n)$. All the other $(n - 1)$-digit words in $x_2, \ldots, x_n, 0, 0, \ldots, 0$ will then be cyclic permutations of $(x_2, \ldots, x_n)$ or will have fewer 1's. Thus, they form a sequence of unmodified nodes, which establishes a connection to $(0, \ldots, 0)$. A similar argument holds for nodes having more 1's than those occurring in $(a_2, \ldots, a_n)$.

Finally, consider $1, \ldots, 1, 0, \ldots, 0$: that is, $n - 1$ 1's followed by $n - 1$ 0's. This provides a connection from $(1, \ldots, 1)$ to $(0, \ldots, 0)$, unless $(a_2, \ldots, a_n)$ is a cyclic permutation of a block of 1's followed by a block of 0's. In the latter event, $1, \ldots, 1, 0, 1, 0, \ldots, 0$ will be a connection, unless $(a_2, \ldots, a_n)$ is a cyclic permutation of $(1, \ldots, 1, 0)$ or $(1, 0, \ldots, 0)$. In the latter case $1, \ldots, 1, 0, 0, 1, 1, 0, \ldots, 0$ has exactly two words, $(1, \ldots, 1, 0)$ and $(1, 0, \ldots; 0)$, which are not nodes in the modified diagram. However, the fact that the sequence of edges given by this sequence is a legitimate connection through the modified graph is a consequence of the fact that $f^1(1, \ldots, 1, 0) = f^1(1, 0, \ldots, 0) = 0$.

It is desirable to have a large class of sequences obtainable from a single shift register by some simple modifications of the feedback. A group theoretic approach would be to have a large collection of operators on feedbacks such that, if any operator is applied to the feedback of any de Bruijn sequence, it yields the feedback of some de Bruijn sequence. Theorem 22 shows that there are only four desirable operators of the type defined by modulo 2 addition of functions to the feedback function, so that this type of modification would not significantly increase the versatility of a shift register.

3. EXPERIMENTAL RESULTS

An experimental program was conducted to determine the distribution of sequence lengths for nonlinear shift register sequences. In each series of tests, a fixed register length and a fixed feedback logic were used, but the selection of taps leading from the shift register into the feedback logic was varied. For simplicity and uniformity, the shift register was started in the *all 1's* state, and the number of clock intervals until the shift register returned to the *all 1's* state was counted on an auxiliary time-interval meter.

3.1.   *Conclusions established by the tests*

The tests furnished valuable information of several types:

1. The hypothesis that the distribution of cycle lengths for nonlinear shift registers is a flat distribution from 1 to 2^n (as adduced in Section 2.1) was substantiated as valid for the type of sampling used.

2. Specific tap combinations yielding near-maximum-length sequences were determined, as a by-product of the length-distribution studies mentioned above.

3. Anomalies of the nonlinear logics in question were observed, including predictable tap combinations yielding extremely short sequences.

4. Sufficient data were obtained to establish conclusively the absence of a simple algebraic theory governing the sequences obtained from the various nonlinear logics. Specifically, the divisibility patterns among the sequence lengths for linear sequences have no analog in the nonlinear case.

5. The flat distribution of sequence lengths was observed with true random statistical fluctuation. There was no pseudorandom biasing to force a complete set of tests (i.e., all possible tap combinations with fixed register length and feedback logic) to be totally flat or to have a mean length of exactly 2^{n-1}.

6. Cases where different tap combinations yielded the same sequence length were singled out for further study. The possibility of a hidden relationship in such cases appears extremely unlikely, since their occurrence was in agreement with the predicted frequency for "accidental equality" based on the flat statistical model.

7. In one case, all cycles corresponding to a given logic and fixed tap combinations were measured. Thus, data were compiled relating to the number as well as to the lengths of shift register cycles.

3.2.   *Nonlinear Logic I*

The nonlinear logic studied most extensively was

$$a_k = w \oplus y \oplus z \oplus xy \oplus xz \oplus yz , \tag{1}$$

henceforth called Logic I. Here $w = a_{k-n}$, and x, y, z are any three distinct members of the set $a_{k-1}, a_{k-2}, \ldots, a_{k-n+1}$. (Henceforth, w, x, y, z also represent the subscripts: thus, $w = n$, etc.)

In view of the symmetric role of y and z, it was no loss of generality to impose the condition $y < z$. Moreover, since the transformation replacing $a_k = f(a_{k-1}, \ldots, a_{k-n+1}) \oplus a_{k-n}$ by $a_k = f(a_{k-n+1}, \ldots, a_{k-1}) \oplus a_{k-n}$, produces the same sequence, but running in reverse order, it was sufficient to consider only the cases $x \geq n/2$. (That is, the cycle length is the same, whether the sequence runs forward or backward.)

Logic I has a balanced truth table, presented in Table 10. The bottom half of the truth table is the complement of the top half, and each half has the same number of 1's and 0's. Thus, Theorems 1, 2, 5, and 7 of Section 2 are applicable.

Table VI-10. Truth table for the logic
$$w \oplus y \oplus z \oplus xy \oplus xz \oplus yz.$$

w	x	y	z	Output
0	0	0	0	0
0	0	0	1	1
0	0	1	0	1
0	0	1	1	1
0	1	0	0	0
0	1	0	1	0
0	1	1	0	0
0	1	1	1	1
1	0	0	0	1
1	0	0	1	0
1	0	1	0	0
1	0	1	1	0
1	1	0	0	1
1	1	0	1	1
1	1	1	0	1
1	1	1	1	0

The shift register was given the initial state *all 1's*. For even n, where n represents the register length, the resulting sequences had minimum length $3n/2$; for odd n, the minimum length was $3n-2$. For even $n \geq 6$, the minimum sequence length is attained by the tap combination

$$w = n, \quad x = 1, \quad y = 2, \quad z = n/2. \tag{2}$$

For odd $n \geq 5$, the minimum sequence length is attained by the tap combination

$$w = n, \quad x = 1, \quad y = 2, \quad z = n - 1. \tag{3}$$

(The tap combination for the shortest sequence length is not, in general, unique.)

These situations are illustrated in Table VI-11 for the cases $n = 5$ (with period $3 \times 5 - 2 = 13$) and $n = 8$ (with period $3 \times 8/2 = 12$). As seen on an oscilloscope, the tap combinations (2) and (3) yield the characteristic shapes shown in Figures VI-12 and VI-13.

The maximum length for sequences using Logic I is $2^n - 1$, which was attained one or more times for many, but not all, values of n. No simple rule has yet been discovered for these maximum-length sequences, and it is probable that no simple rule exists.

Section 4.1 shows the periods obtained for $n = 11$, using Logic I, the initial state *all 1*'s, and all possible tap combinations. The periods are presented in numerical order (Table VI-13), and a histogram (Fig. VI-15)

Table VI-11. Minimal sequences of degree 5 and degree 8
using the feedback logic
$$w \oplus y \oplus z \oplus xy \oplus xz \oplus yz$$
and the initial register state *all ones*.

w	z	—	y	x	w	—	—	—	y	—	z	x
1	1	1	1	1	1	1	1	1	1	1	1	1
1	1	1	1	0	1	1	1	1	1	1	1	0
1	1	1	0	0	1	1	1	1	1	1	0	0
1	1	0	0	0	1	1	1	1	1	0	0	0
1	0	0	0	0	1	1	1	1	0	0	0	0
0	0	0	0	1	1	1	1	0	0	0	0	1
0	0	0	1	0	1	1	0	0	0	0	1	1
0	0	1	0	1	1	0	0	0	0	1	1	1
0	1	0	1	0	0	0	0	0	1	1	1	1
1	0	1	0	1	0	0	0	1	1	1	1	1
0	1	0	1	1	0	0	1	1	1	1	1	1
1	0	1	1	1	0	1	1	1	1	1	1	1
0	1	1	1	1								

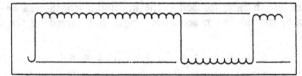

Fig. VI-12. Sequence for tap combination (2) with
even register length $n = 20$
(period $(3/2) \times 20 = 30$).

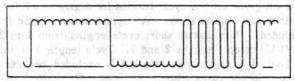

Fig. VI-13.   Sequence for tap combination (3) with
odd register length $n = 11$
(period $3 \times 11 - 2 = 31$).

indicates the qualitative nature of the distribution of sequence lengths.

Section 4.2 shows the periods obtained for $n = 13$, using Logic I, the initial state *all 1's*, and all possible tap combinations (Table VI-14 and Fig. VI-16).

Section 4.3 presents a sampling of the periods obtained for $n = 20$, using Logic I, the initial state *all 1's*, and randomly selected tap positions (Table VI-15 and Fig. VI-17). (Cases in which w, x, y, z had a common factor were excluded.)

Section 4.4 presents a similar sampling for $n = 31$ (Table VI-16 and Fig. VI-18), followed by an enumeration of some of the sequences of degree 31 within 2 per cent of maximum length (Table VI-17).

Table VI-12.   Truth table for the logic
$1 \oplus w \oplus y \oplus z \oplus xy \oplus yz$.

w	x	y	z	Output
0	0	0	0	1
0	0	0	1	0
0	0	1	0	0
0	0	1	1	0
0	1	0	0	1
0	1	0	1	0
0	1	1	0	1
0	1	1	1	1
1	0	0	0	0
1	0	0	1	1
1	0	1	0	1
1	0	1	1	1
1	1	0	0	0
1	1	0	1	1
1	1	1	0	0
1	1	1	1	0

Section 4.5 gives all the cycle lengths for degree $n = 7$ for each of the 30 distinct tap combinations. A representative vector from each cycle is included. Note that all short cycle lengths (from 1 to 20) occur in Table VI-18 except lengths 2 and 7. Cycle length 2 is impossible whenever Logic I is used; cycle length 7 is excluded because 7 is the register length in the cases tested.

3.3. *Nonlinear Logic II*

Another nonlinear logic used in the test program was

$$1 \oplus w \oplus y \oplus z \oplus xy \oplus yz , \qquad (4)$$

henceforth called Logic II, which has the truth table shown in Table VI-12.

Theorems 1, 2, 5, and 7 of Section 2 are again applicable. Now, however, $2^n - 2$ is an upper bound on the sequence length obtainable, since there must be an even number of cycles (which rules out 2^n as a possible length), and neither $00...0$ nor $11...1$ will form a single-element cycle.

The symmetry between y and z, which was observed to hold for Logic I, is no longer present when Logic II is used. However, the

w	x	y	z	Period	Sequence
11	10	1	6	20	
11	7	4	10	15	
8	5	3	7	11	

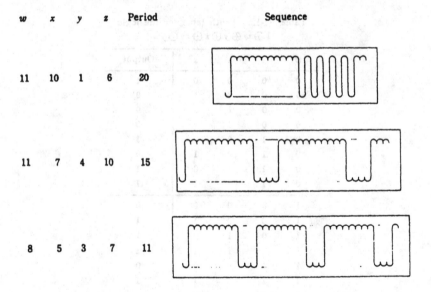

Fig. VI-14. Three short cycles obtained using Logic II.

operation of replacing x, y, and z by $n - x, n - y$, and $n - z$, respectively, still preserves cycle lengths.

As for Logic I, it is not difficult to specify the shortest cycle lengths obtainable from Logic II, where the *all 1's* vector has been used as the starting point. It is also possible to describe tap combinations yielding these shortest cycles.

If n is even, a cycle of length $(3n/2) - 1$ is obtained by letting $w = n$, $x = (n/2) + 1$, $y = (n/2) - 1$, and $z = n - 1$. If n is odd, a cycle of length $(3/2)(n - 1)$ is obtained by letting $w = n$, $x = (1/2)(n + 3)$, $y = (1/2)(n - 3)$, and $z = n - 1$. A "next-shortest" cycle for n odd, of length $2n - 2$, is obtained by letting $w = n$, $x = n - 1$, $y = 1$, and $z = (1/2)(n + 1)$. These three situations are illustrated in Figure VI-14. Section 4.6 shows the set of periods obtained using Logic II with $n = 8$, and Section 4.7 shows the periods with $n = 11$. The starting vector in both cases was the *all 1's* vector. The data are presented in several forms (Tables VI-19 and VI-20, Figs. VI-19 and VI-20), as was done for Logic I.

4. TABULATION OF EXPERIMENTAL DATA

4.1. *Distribution of periods for Logic I and $n = 11$*

This section shows the periods obtained for $n = 11$, using Logic I, the initial state *all 1's*, and all possible tap combinations for x, y, and z. The periods are presented in numerical order in Table VI-13, and their distribution is portrayed in a histogram (Fig. VI-15).

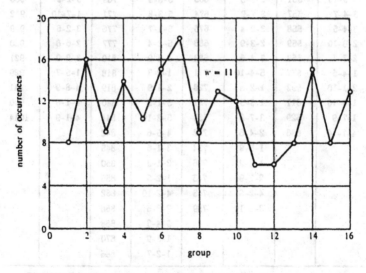

Fig. VI-15. Distribution of periods into sixteenths of the interval from 1 to 2^{11} (Logic I, $n = 11$).

Table VI-13. Periods obtained for all tap combinations (Logic I, $n = 11$).

Tap Combination	Cycle Length	Tap Combination	Cycle Length	Tap Combination	Cycle Length	Tap Combination	Cycle Length
Group 1 1 to 127		Group 2 128 to 255		Group 3 256 to 383		Group 4 384 to 511	
1-2-3	31	4-2-8	152	1-7-8	288	5-1-4	389
4-1-3	47	3-1-2	161	3-7-10	308	4-6-8	391
5-4-8	62	2-1-5	165	2-1-7	312	2-8-9	392
2-4-8	66	5-6-10	174	2-3-10	324	1-5-8	395
4-7-9	69	1-2-9	177	4-2-6	356	3-9-10	396
3-2-8	79	3-1-4	182	2-6-10	358	3-2-10	412
1-4-7	114	1-3-4	186	4-1-10	360	3-4-8	426
2-5-8	117	4-1-7	186	1-6-8	364	3-1-5	436
		1-9-10	191	4-8-10	369	5-1-3	456
		3-2-5	198			4-6-10	458
		4-3-10	205			2-5-6	464
		2-9-10	216			5-1-10	471
		1-3-8	225			1-3-7	484
		2-4-5	233			5-1-6	486
		2-5-7	235				
		1-3-9	244				
Group 5 512 to 639		Group 6 640 to 767		Group 7 768 to 895		Group 8 896 to 1023	
5-9-10	531	2-7-8	663	5-8-9	769	5-2-9	900
3-4-7	557	3-7-9	674	5-3-8	771	1-7-10	912
3-4-5	558	3-2-4	675	5-1-7	776	3-2-6	919
2-8-10	559	2-3-9	683	5-2-4	777	2-6-8	920
2-3-7	568	3-4-6	683	1-4-9	810	3-2-9	921
1-4-5	577	5-4-10	711	1-2-3	818	1-5-7	929
2-1-10	592	1-2-8	728	2-1-9	819	4-8-9	941
1-6-10	611	1-3-5	731	5-1-8	840	4-1-6	990
1-7-9	629	3-7-8	731	5-2-10	841	4-1-9	1014
4-7-10	636	2-4-6	732	4-5-6	842		
		1-6-9	734	3-1-6	845		
		1-8-9	736	2-1-3	850		
		2-7-9	745	1-2-5	855		
		4-2-5	755	4-2-10	862		
		3-6-10	759	2-3-5	868		
				5-4-6	868		
				1-5-9	870		
				1-2-7	886		

Table VI-13 (Cont'd).

Tap Combination	Cycle Length	Tap Combination	Cycle Length	Tap Combination	Cycle Length	Tap Combination	Cycle Length
Group 9 1024 to 1151		Group 10 1152 to 1279		Group 11 1280 to 1407		Group 12 1408 to 1535	
3-1-10	1049	4-3-8	1154	1-4-10	1338	5-4-7	1409
1-8-10	1051	1-4-8	1156	3-1-9	1343	5-6-7	1418
3-5-6	1051	5-3-10	*1174	1-6-7	1349	5-6-8	1459
4-1-5	1064	2-3-4	1182	4-7-8	1358	3-5-10	1487
3-2-7	1070	3-4-9	1184	1-3-10	1380	4-3-9	1522
5-2-3	1075	3-1-7	1185	3-8-9	1385	5-6-9	1530
4-3-7	1095	5-4-9	1218				
3-5-9	1096	5-7-10	1236				
5-3-9	1107	4-1-8	1237				
4-1-2	1108	2-4-9	1247				
1-5-10	1110	5-1-9	1252				
3-8-10	1111	5-2-7	1264				
4-2-9	1130						
Group 13 1536 to 1663		Group 14 1664 to 1791		Group 15 1792 to 1919		Group 16 1920 to 2047	
3-6-8	1589	2-3-8	1664	5-7-8	1798	1-2-4	1934
5-3-6	1599	4-6-7	1664	2-1-6	1829	5-7-9	1956
3-4-10	1605	4-5-10	1668	3-6-9	1856	2-5-10	1965
1-3-6	1609	4-5-9	1676	2-4-7	1862	5-2-8	1971
4-2-7	1622	3-1-8	1684	3-6-7	1878	2-6-7	1972
1-5-6	1639	4-2-3	1692	4-3-6	1888	4-6-9	1976
1-2-6	1650	4-5-7	1693	1-4-6	1906	2-5-9	1986
4-3-5	1659	3-5-7	1713	2-7-10	1908	5-3-4	1993
		5-1-2	1717			2-1-8	2009
		5-3-7	1733			5-8-10	2023
		4-9-10	1753			2-4-10	2030
		2-3-6	1756			3-5-8	2047
		4-5-8	1757			5-2-6	2047
		2-1-4	1765				
		2-6-9	1769				

Average cycle length = 988.194

$2^{n-1} = 1024$

4.2.   *Distribution of periods for Logic I and $n = 13$*

This section shows the periods obtained for $n = 13$, using Logic I, the initial state *all 1's*, and all possible tap combinations for x, y, and z. The periods are presented in numerical order in Table VI-14, and their distribution is shown in Figure VI-16.

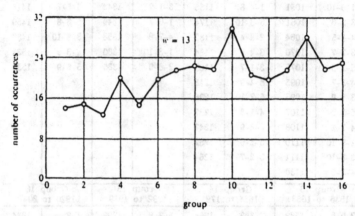

Fig. VI-16.   **Distribution of periods into sixteenths of the interval from 1 to 2^{13} (Logic I, $n = 13$).**

Table VI-14. Periods obtained for all tap combinations (Logic I, $n = 13$).

Tap Combination	Cycle Length	Tap Combination	Cycle Length	Tap Combination	Cycle Length	Tap Combination	Cycle Length
Group 1 1 to 511		Group 2 512 to 1023		Group 3 1024 to 1535		Group 4 1536 to 2047	
1-2-12	37	1-4-8	516	2-5-12	1046	3-9-11	1536
4-2-12	37	1-10-11	531	3-4-12	1074	2-1-12	1542
2-8-12	261	1-2-4	546	2-7-11	1105	2-9-10	1580
1-5-9	268	5-1-10	557	1-6-12	1106	5-2-11	1605
1-6-9	269	6-2-9	588	4-5-8	1128	1-3-10	1617
1-9-12	287	6-4-11	622	6-4-9	1153	4-2-5	1620
4-9-10	289	4-9-12	641	6-1-2	1157	6-3-12	1646
1-7-12	316	1-2-7	657	1-2-8	1309	1-7-11	1663
5-6-11	335	5-2-8	660	4-2-11	1317	2-1-7	1712
4-1-8	400	6-1-10	731	6-5-9	1375	1-3-12	1714
4-1-3	427	2-5-8	759	3-6-8	1416	6-5-12	1757
2-6-12	451	2-3-6	831	5-7-11	1502	2-4-8	1773
4-1-11	466	2-1-8	844	2-3-5	1522	5-8-11	1788
6-7-12	486	6-3-11	912			2-4-9	1865
		1-9-11	968			4-3-10	1885
						5-3-10	1898
						4-2-6	1913
						4-1-7	2006
						1-6-8	2019
						3-1-4	2047

Tap Combination	Cycle Length	Tap Combination	Cycle Length	Tap Combination	Cycle Length	Tap Combination	Cycle Length
Group 5 2048 to 2559		Group 6 2560 to 3071		Group 7 3072 to 3583		Group 8 3584 to 4095	
3-8-11	2138	3-1-7	2594	1-3-11	3095	6-10-12	3611
4-7-12	2178	1-4-10	2663	6-5-7	3107	5-1-4	3626
1-2-3	2197	2-3-8	2665	2-4-6	3166	5-8-12	3631
2-7-8	2202	4-5-9	2673	1-6-10	3180	6-9-10	3654
4-10-12	2215	3-7-8	2702	6-3-4	3187	4-2-9	3692
3-9-10	2227	2-4-7	2719	3-6-12	3214	3-6-7	3711
3-7-11	2247	5-3-9	2751	6-7-8	3256	6-4-10	3813
2-6-10	2326	3-1-6	2795	3-1-12	3315	2-7-10	3847
6-8-11	2350	5-3-6	2816	3-2-7	3317	3-4-8	3852
5-7-12	2360	6-2-11	2822	6-9-11	3322	4-3-9	3873
1-4-6	2389	5-7-9	2823	3-1-8	3329	6-4-5	3900
2-1-4	2389	5-9-12	2840	2-5-11	3360	1-4-5	3903
5-3-7	2541	1-2-6	2894	1-5-6	3368	4-7-11	3918
6-3-9	2545	3-11-12	2936	2-5-10	3389	3-6-9	3957

Table VI-14 (Cont'd).

Tap Combination	Cycle Length	Tap Combination	Cycle Length	Tap Combination	Cycle Length	Tap Combination	Cycle Length
5-10-12	2549	1-11-12	2956	1-9-10	3426	2-1-5	3987
		4-1-12	2978	2-9-12	3428	4-5-11	3998
		4-3-5	2993	2-3-12	3476	4-7-9	4009
		5-1-7	2993	5-1-6	3483	6-11-12	4025
		3-6-10	3004	1-8-12	3510	5-2-3	4043
		2-1-3	3046	4-1-10	3518	6-2-7	4061
				4-3-6	3529	2-5-9	4062
				1-3-4	3550	2-4-5	4073
						4-10-11	4078

Group 9 4096 to 4607		Group 10 4608 to 5119		Group 11 5120 to 5631		Group 12 5632 to 6143	
6-1-7	4101	3-2-8	4621	6-4-12	5147	1-2-11	5635
2-1-6	4160	6-1-12	4651	1-3-9	5157	1-3-6	5656
6-3-7	4167	2-3-9	4661	3-4-11	5191	2-1-9	5666
6-1-9	4170	4-7-10	4672	1-5-11	5196	5-7-8	5688
5-3-11	4206	1-3-8	4689	2-6-11	5207	4-2-3	5704
2-10-12	4247	4-3-8	4707	6-2-8	5258	4-2-7	5718
4-9-11	4259	6-7-10	4710	4-5-7	5282	6-3-5	5720
3-8-9	4262	1-2-5	4712	2-11-12	5300	3-1-9	5754
4-3-12	4290	5-4-8	4719	1-7-10	5317	3-2-10	5760
1-4-11	4306	6-7-9	4751	6-10-11	5339	3-10-12	5781
1-6-11	4332	3-1-2	4795	5-2-10	5349	5-6-9	5840
2-7-9	4354	5-3-4	4799	1-6-7	5380	5-1-2	5857
4-8-11	4385	5-9-10	4819	5-3-12	5387	5-2-6	5885
1-5-7	4409	4-8-10	4845	1-4-12	5390	4-1-6	5895
1-8-10	4483	5-1-12	4853	3-2-4	5410	2-8-10	5936
2-7-12	4485	2-8-11	4863	3-4-7	5453	2-4-10	6003
1-5-8	4493	4-6-10	4865	6-3-10	5492	5-6-7	6096
4-6-9	4532	3-2-11	4874	1-4-7	5558	6-8-12	6107
5-1-9	4535	5-4-12	4875	6-1-3	5590	1-2-9	6124
3-4-9	4539	6-2-3	4910	2-9-11	5595	5-4-10	6133
1-10-12	4544	6-5-11	4916	6-9-12	5613		
6-2-10	4549	3-4-5	4918				
		6-5-8	4943				
		2-1-11	4954				
		5-9-11	4973				
		5-2-7	4990				
		2-3-4	5013				

Table VI-14 (Cont'd).

Tap Combination	Cycle Length	Tap Combination	Cycle Length	Tap Combination	Cycle Length	Tap Combination	Cycle Length
		1-4-9	5044				
		3-1-11	5055				
		6-4-8	5110				
Group 13 6144 to 6655		Group 14 6656 to 7167		Group 15 7168 to 7679		Group 16 7680 to 8191	
3-4-10	6148	5-4-7	6668	3-7-9	7204	3-2-9	7702
2-3-10	6160	2-3-7	6677	6-8-10	7205	3-5-10	7725
3-5-7	6164	6-3-8	6701	3-5-6	7222	5-1-11	7734
2-5-6	6180	5-6-10	6705	6-1-5	7259	3-6-11	7748
5-7-10	6188	1-2-10	6724	5-8-9	7337	4-1-2	7783
6-7-11	6201	3-4-6	6736	3-8-10	7343	4-1-9	7829
4-6-11	6274	4-7-8	6740	4-3-7	7365	2-6-9	7859
1-8-11	6277	3-5-11	6783	1-7-9	7388	5-2-12	7874
4-8-12	6300	3-4-8	6789	2-6-7	7388	1-3-5	7880
4-1-5	6319	3-2-5	6793	4-3-11	7434	5-2-9	7888
5-1-3	6322	5-6-8	6814	5-4-6	7438	4-6-7	7891
6-1-4	6332	4-2-10	6826	5-11-12	7445	4-11-12	7923
3-10-11	6447	3-9-12	6830	5-1-8	7480	4-8-9	7936
3-8-12	6470	1-3-7	6872	2-6-8	7498	5-4-9	7950
4-5-6	6498	2-3-11	6949	1-8-9	7507	5-2-4	8006
6-1-11	6515	1-5-6	6977	4-6-8	7548	3-7-12	8017
5-8-10	6516	3-5-9	6980	3-2-12	7549	3-1-5	8022
5-6-12	6541	3-7-10	6983	5-10-11	7577	2-10-11	8083
5-3-8	6564	4-2-8	7015	6-8-9	7627	2-5-7	8099
6-5-10	6599	2-1-10	7054	4-5-10	7645	3-2-4	8110
2-8-9	6616	6-2-12	7075	6-2-4	7648	5-4-11	8144
6-2-5	6630	6-1-8	7080	6-4-7	7668	2-4-11	8158
		4-5-12	7093			3-1-10	8191
		3-5-12	7135				
		2-4-12	7146				
		4-6-12	7148				
		1-5-10	7156				
		1-7-8	7167				

Average cycle length = 4458.615

$2^{n-1} = 4096$

4.3.   *Distribution of periods for Logic I and n = 20*

This section presents a sampling of the periods obtained for $n = 20$, using Logic I, the initial state *all 1's*, and 100 randomly selected tap positions. (Cases in which $(w, x, y, z) \geq 1$ were excluded.)   The periods are listed in numerical order in Table VI-15, and their distribution appears in Figure VI-17.

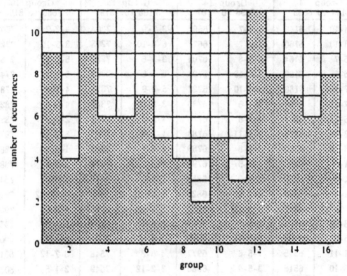

Fig. VI-17.   Distribution of periods into sixteenths of the interval from 1 to 2^{20} (Logic I, $n = 20$).

Table VI-15. Periods obtained for 100 tap combinations
(Logic I, $n = 20$).

Tap Combination	Cycle Length	Tap Combination	Cycle Length	Tap Combination	Cycle Length	Tap Combination	Cycle Length
Group 1 1 to 65,535		Group 2 65,536 to 131,071		Group 3 131,072 to 196,607		Group 4 196,608 to 262,143	
8-4-5	86	6-3-13	67,851	3-2-7	160,271	7-1-13	220,890
10-5-6	120	1-2-14	73,940	7-2-16	169,356	2-9-16	225,127
10-8-9	124	9-7-13	90,261	1-13-17	174,095	4-3-10	238,983
8-9-12	126	9-15-19	98,063	5-7-15	175,059	2-3-6	250,854
7-13-19	4,116			1-5-19	176,779	4-15-19	256,185
5-1-2	4,178			7-3-8	183,643	9-1-19	256,422
9-8-11	17,390			8-1-11	183,901		
10-3-7	44,725			3-4-7	187,363		
8-1-4	53,893			9-4-7	188,688		
Group 5 262,144 to 327,679		Group 6 327,680 to 393,215		Group 7 393,216 to 458,751		Group 8 458,752 to 524,287	
1-10-17	264,916	2-5-9	328,773	1-4-7	409,116	4-1-6	474,003
1-9-17	277,817	9-16-18	347,570	9-4-18	418,938	4-5-11	501,072
3-11-12	284,592	5-8-15	358,597	2-11-18	425,067	3-1-13	513,222
6-11-19	294,529	7-4-14	362,633	4-18-19	452,898	10-4-5	522,048
2-12-17	297,036	10-2-17	363,166	5-4-11	455,905		
9-12-17	309,805	1-2-13	366,046				
		4-10-15	387,824				
Group 9 524,288 to 589,823		Group 10 589,824 to 655,359		Group 11 655,360 to 720,895		Group 12 720,896 to 786,431	
8-7-19	536,849	7-1-10	624,822	4-5-12	660,630	4-17-18	747,283
6-7-11	560,082	3-5-15	625,416	2-1-12	662,275	1-16-18	747,784
		4-6-15	633,825	3-6-18	691,316	5-12-14	751,207
		9-1-18	634,341			8-11-15	758,268
		10-5-12	652,803			7-1-3	758,820
						8-9-16	759,525
						5-8-9	765,856
						8-1-5	769,960
						4-9-18	773,522
						2-9-13	777,864
						2-10-19	783,889

Table VI-15 (Cont'd).

Tap Combination	Cycle Length	Tap Combination	Cycle Length	Tap Combination	Cycle Length	Tap Combination	Cycle Length
Group 13 786,432 to 851,967		Group 14 851,968 to 917,503		Group 15 917,504 to 983,039		Group 16 983,040 to 1,048,575	
7-2-15	788,950	1-3-13	855,126	4-3-18	921,674	6-4-9	987,044
2-3-8	794,178	8-18-19	868,466	7-12-19	931,983	5-6-18	988,077
8-7-9	800,182	4-15-16	871,545	3-12-17	935,405	10-7-8	998,020
7-5-17	800,865	9-11-19	880,787	2-12-13	946,350	5-4-13	1,014,910
1-4-6	815,387	1-10-12	890,832	6-3-5	951,268	8-7-14	1,015,417
9-3-13	822,780	9-4-10	897,229	6-3-11	961,385	7-3-6	1,020,778
3-1-18	828,201	5-3-14	914,715			5-1-17	1,025,826
8-4-11	847,437					10-3-14	1,032,357

Average cycle length = 530,016

$2^{n-1} = 524,288$

4.4. *Distribution of periods for Logic I and $n = 31$*

Section 4.4 presents a sampling of the periods obtained for $n = 31$, using Logic I, the initial state *all 1's*, and 100 randomly selected tap positions. The periods are listed in numerical order in Table VI-16, and their distribution is seen in Figure VI-18. An enumeration of some of the sequences of degree 31 within 2 per cent of maximum length appears in Table VI-17.

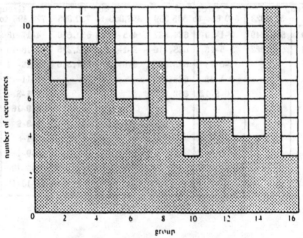

Fig. VI-18.   Distribution of periods into sixteenths of the interval from 1 to 2^{31} (Logic I, $n = 31$).

Table VI-16. Periods obtained for 100 tap combinations (Logic 1, $n = 31$).

Tap Combination	Cycle Length	Tap Combination	Cycle Length	Tap Combination	Cycle Length	Tap Combination	Cycle Length
Group 1 (1 to 134,217,727)		Group 2 (134,217,728 to 268,435,455)		Group 3 (268,435,456 to 402,653,183)		Group 4 (402,653,184 to 536,870,911)	
14-18-19	5,224,868	14-20-29	141,482,660	2-5-29	287,779,860	2-3 20	435,250,982
5-8-20	8,441,967	7-4-12	144,118,009	14-1-19	297,400,661	14 8 21	439,413,230
11-7-24	14,878,669	2-6-30	155,392,865	7-1-2	316,401,877	12 1 21	439,511,522
2-4-20	30,206,942	11-12-28	213,540,639	1-28-29	347,260,128	2-4-24	452,760,129
7-13-23	36,027,873	2-5-16	244,307,667	8-7-10	366,606,954	11-5-30	467,128,458
9-21-29	82,229,293	6-4-30	260,620,416	5-3-18	369,897,479	9-5 25	478,825,614
7-17-20	100,778,549	12-7-10	255,196,102			9-11-18	485,118,365

Tap Combination	Cycle Length	Tap Combination	Cycle Length	Tap Combination	Cycle Length	Tap Combination	Cycle Length
Group 5 (536,870,912 to 671,088,639)		Group 6 (671,088,640 to 805,306,367)		Group 7 (805,306,368 to 939,524,095)		Group 8 (939,524,096 to 1,073,741,823)	
9-5-13	562,206,087	4-11-26	707,054,026	13-7-30	854,321,197	10-3-15	974,103,101
7-12-23	596,751,700	10-15-27	709,880,605	12-3-30	906,758,029	10-3-5	986,092,475
7-10-21	603,606,929	12-15-25	760,486,347	7-9 25	910,345,145	1-24-26	1,003,685,155
10-11-26	607,647,917	4-10-21	762,443,814	12-16-17	915,206,335	8-5-30	1,004,074,445
1-26-28	613,257,912	13-1-21	787,930,917	3-11-28	935,539,443	7-6-19	1,007,604,040
13-8-29	615,196,281	13-5-9	797,827,012			12-9 14	1,033,192,158
6-8-24	633,446,247					7-13-17	1,049,849,426
12-16-25	635,637,611					10-6-24	1,050,466,605
6-12-24	649,706,666						
13-17-23	659,179,867						

Table VI-16 (Cont'd).

Tap Combination	Cycle Length	Tap Combination	Cycle Length	Tap Combination	Cycle Length	Tap Combination	Cycle Length
Group 9 1,073,741,824 to 1,207,959,551		Group 10 1,207,959,552 to 1,342,177,279		Group 11 1,342,177,280 to 1,476,395,007		Group 12 1,476,395,008 to 1,610,612,735	
3-11-24	1,078,896,621	4-16-28	1,307,691,468	4-6-28	1,399,305,881	11-10-25	1,481,183,201
6-2-16	1,108,233,683	1-14-17	1,341,744,315	10-11-13	1,404,678,625	8-3-10	1,502,437,007
13-11-20	1,115,994,733	13-27-29	1,341,795,279	9-28-29	1,407,831,161	12-22-24	1,511,160,969
6-9-12	1,132,906,904			7-18-27	1,420,510,772	11-4-15	1,518,457,561
5-22-25	1,148,674,456			8-18-25	1,450,248,144	11-20-30	1,578,986,926

Tap Combination	Cycle Length	Tap Combination	Cycle Length	Tap Combination	Cycle Length	Tap Combination	Cycle Length
Group 13 1,610,612,736 to 1,744,830,463		Group 14 1,744,830,464 to 1,879,048,191		Group 15 1,879,048,192 to 2,013,265,919		Group 16 2,013,265,920 to 2,147,483,647	
14-2-24	1,642,076,260	9-8-22	1,771,274,210	9-20-22	1,881,753,856	12-1-20	2,111,380,835
2-17-27	1,693,694,852	10-22-28	1,790,197,261	12-4-14	1,935,497,848	7-10-14	2,111,834,786
1-18-21	1,713,700,520	7-4-19	1,858,744,433	12-15-24	1,952,149,928	8-6-21	2,146,670,359
8-16-30	1,735,705,692	14-15-20	1,860,633,402	14-1-9	1,954,336,898		
				14-4-10	1,964,932,265		
				8-15-26	1,976,292,819		
				10-14-20	1,977,808,372		
				4-12-13	1,982,179,928		
				15-18-27	1,986,204,013		
				11-8-19	1,987,986,835		
				1-10-22	1,994,233,668		

Average cycle length = 977,144,051

$2^{n-1} = 1,073,741,824$

Table VI-17. A selection of sequences within 2 per cent of
maximum length (Logic I, $n = 31$).

x	y	z	Cycle Length
1	2	14	2,127,557,162
1	2	28	2,145,180,558
1	3	13	2,123,236,183
1	3	21	2,121,731,272
1	3	27	2,122,884,589
3	11	25	2,130,648,058
6	7	20	2,128,192,327
7	10	11	2,111,834,786
8	6	21	2,146,670,359
12	1	20	2,111,380,835

4.5. *Complete decomposition into cycles for Logic I and $n = 7$*

This section presents the complete cycle decomposition for shift
registers of degree 7, using Logic I and all of the 30 essentially distinct
tap combinations for x, y, and z. A representative vector from each
cycle is included. Note that all short cycle lengths (from 1 to 20) occur
in Table VI-18 with the exception of lengths 2 and 7. Also, the total
number of cycles for each tap position is even.

Table VI-18. Complete cycle decomposition (Logic 1, $n = 7$).

x-y-z	V_1	P_1	V_2	P_2	V_3	P_3	V_4	P_4	V_5	P_5	V_6	P_6	V_7	P_7	V_8	P_8
1-2-3	1111111	127	0000000	1	0110110	3	0000000	1								
1-2-4	1111111	90	0000001	34	1001001	8	0001000	5	0110110	3	0000000	1				
1-2-5	1111111	99	0101111	12	0000111	22	0010100	1	0001000	4	0000000	1				
1-2-6	1111111	19	1110111	86	1110111	11	0010111	5								
1-3-4	1111111	81	1001001	26	0011101	10	0000000	1	1110111	11	0000000	1				
1-3-5	1111111	106	1110111	11	0101011	30	0001011	20								
1-3-6	1111111	27	1001011	39	0101101	3	0000001	1								
1-4-5	1111111	115	1000110	9	0110010	22	0000000	1								
1-4-6	1111111	79	1100011	26	1110111	33	0000000	1								
1-5-6	1111111	32	1110111	62	0101111											
2-1-3	1111111	17	1110111	58	1100111	39	1000001	10	1001001	3	0000000	1				
2-1-4	1111111	127	0000000	1												
2-1-5	1111111	104	1110111	15	0110000	8	1011100	13	0110110	8	0001000	6	1001001	3	0000000	1
2-1-6	1111111	52	1110111	32	1101111	13	0000000	1								
2-3-4	1111111	120	0001000	4	1001001	3	0000000	1	1001001	3	0000000	1				
2-3-5	1111111	37	1110111	51	0101111	39	0101111	1								
2-3-6	1111111	101	1001000	14	0001011	12	0000000	1								
2-4-5	1111111	127	0000000	1												
2-4-6	1111111	37	1110111	56	0111110	16	0011100	15								
2-5-6	1111111	127	0000000	1												
3-1-2	1111111	113	1001111	10	1110111	4	0000000	1								
3-1-4	1111111	127	0000000	1												
3-1-5	1111111	68	0101111	37	1011111	22	0000000	6	1110111	4	0000000	1				
3-1-6	1111111	71	0001111	33	1100111	13	0001000	1								
3-2-4	1111111	18	1110111	101	0110111	8	0000000	1	0110110	3	0000000	1				
3-2-5	1111111	53	1011111	62	0001000	5	1110111	1								
3-2-6	1111111	79	1110111	45	0110110	3	0000000	1								
3-4-5	1111111	77	1110111	42	0010011	8	0000000	1								
3-4-6	1111111	113	0101111	8	1001011	6	0000000	1								
3-5-6	1111111	52	0001111	54	0010100	14	1110111	4	0110110	3	0000000	1				

4.6. *Distribution of periods for Logic II and n = 8*

Section 4.6 presents the set of periods obtained for $n = 8$, using Logic II, the initial state *all 1's*, and all possible tap combinations. The results are presented in numerical order (Table VI-19) and in histogram form (Fig. VI-19).

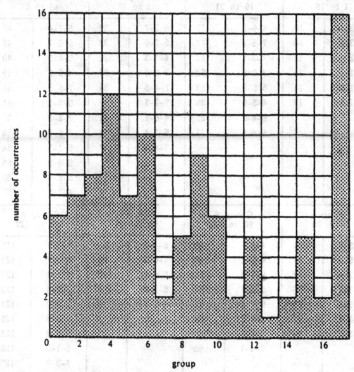

Fig. VI-19. Distribution of periods into sixteenths of the interval from 1 to 2^8 (Logic II, $n = 8$).

Table VI-19. Periods obtained for all tap combinations
(Logic II, $n = 8$).

Tap Combination	Cycle Length	Tap Combination	Cycle Length	Tap Combination	Cycle Length	Tap Combination	Cycle Length
Group 1 1 to 15		Group 2 16 to 31		Group 3 32 to 47		Group 4 48 to 63	
5-3-7	10	4-6-2	17	6-2-7	32	5-7-4	48
5-3-6	10	7-5-1	21	5-7-6	32	5-6-1	49
6-2-4	13	7-3-1	21	4-7-3	32	5-4-3	49
7-1-4	14	7-3-5	22	7-2-6	33	4-5-3	49
7-1-2	14	6-5-4	25	7-4-5	38	7-5-6	50
7-1-6	15	4-5-6	25	7-2-5	39	5-3-1	51
		6-4-2	27	4-6-1	41	7-2-3	52
		6-4-5	31	5-1-4	46	7-5-3	52
				5-4-7	47	5-1-3	54
						5-6-2	55
						5-6-4	62
						6-2-5	63
Group 5 64 to 79		Group 6 80 to 95		Group 7 96 to 111		Group 8 112 to 127	
5-3-2	67	4-7-1	80	6-1-3	98	6-5-7	112
6-5-3	67	4-5-1	80	4-6-5	99	6-4-7	120
7-2-4	72	5-7-3	81	5-6-3	107	7-4-6	121
5-1-6	75	5-2-7	81	5-2-3	109	7-2-6	123
6-1-5	75	6-7-4	81	6-1-4	110	5-7-2	124
6-3-2	77	5-4-2	84			6-3-5	125
6-5-2	77	6-4-3	84			6-7-2	126
		7-4-1	86			6-1-2	126
						7-3-6	127
Group 9 128 to 143		Group 10 144 to 159		Group 11 160 to 175		Group 12 176 to 191	
4-6-7	131	6-1-7	147	6-7-5	164	6-2-3	180
7-6-4	132	4-5-7	149	4-7-5	164	6-7-1	181
4-7-2	139	7-5-4	150	4-7-6	168		
7-1-5	141			5-3-4	171		
5-1-7	141			6-2-1	173		
				7-6-2	174		

Table VI-19 (Cont'd).

Tap Combination	Cycle Length	Tap Combination	Cycle Length	Tap Combination	Cycle Length	Tap Combination	Cycle Length
Group 13 192 to 207		Group 14 208 to 223		Group 15 224 to 239		Group 16 240 to 255	
6-3-4	196	7-1-3	220	5-4-1	224	5-2-1	241
4-5-2	196	5-7-1	220	7-4-3	225	7-6-3	241
		5-4-6	223	6-7-3	227	6-5-1	248
				5-1-2	227	7-3-2	249
						6-4-1	252
						6-3-7	252
						6-3-1	252
						7-6-2	253
						7-4-2	253
						7-3-6	253
						5-6-7	253
						5-2-4	253
						4-6-3	253
						7-6-5	254
						7-6-1	254
						7-2-1	254

Average cycle length = 119.15
$2^{n-1} = 128$

4.7. *Distribution of periods for Logic II and $n = 11$*

This section presents the set of periods obtained for $n = 11$, using Logic II, the initial state *all 1's*, and all possible tap combinations. These results appear in numerical order in Table VI-20 and graphically in Figure VI-20.

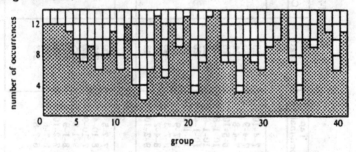

Fig. VI-20. Distribution of periods into intervals of length 50 (Logic II, $n = 11$).

Table VI-20.　Periods obtained for all tap combinations (Logic II, $n = 11$).

Group 1 1 to 49		Group 2 50 to 99		Group 3 100 to 149		Group 4 150 to 199	
Tap Combination	Cycle Length	Tap Combination	Cycle Length	Tap Combination	Cycle Length	Tap Combination	Cycle Length
7-4-10	15	8-3-6	54	7-3-4	105	9-7-6	154
7-4-9	15	7-9-6	56	9-5-3	109	9-4-7	164
7-4-8	15	6-5-3	58	8-6-2	110	6-5-1	166
6-5-10	16	8-6-5	59	6-9-1	119	10-1-9	167
10-1-8	20	7-9-10	61	10-4-7	123	9-6-7	167
10-1-6	20	10-9-7	61	8-7-1	128	10-3-4	169
10-1-4	20	10-3-4	66	9-8-5	129	7-2-3	193
10-1-2	20	8-2-5	66	9-3-2	130	7-1-10	196
8-7-9	40	9-2-8	76	9-8-2	130	9-5-10	197
9-7-3	45	9-2-4	76	7-5-9	132	7-3-1	198
8-4-2	45	7-9-3	76	10-4-8	140	10-3-9	199
9-2-10	48	6-10-9	91	9-6-2	146		

Group 5 200 to 249		Group 6 250 to 299		Group 7 300 to 349		Group 8 350 to 399	
Tap Combination	Cycle Length	Tap Combination	Cycle Length	Tap Combination	Cycle Length	Tap Combination	Cycle Length
7-2-5	209	6-1-3	257	6-4-1	302	9-4-8	354
8-10-4	212	6-8-4	268	10-7-5	303	8-4-9	354
7-10-9	220	7-1-5	273	8-3-2	303	7-3-6	372
9-1-8	225	10-2-9	274	9-8-3	304	6-3-7	372
8-1-9	225	6-7-10	276	9-10-5	305	10-3-9	392
7-2-10	228	10-7-6	277	10-5-8	323	8-5-1	393
8-7-6	243	6-4-8	289	8-2-4	323		
6-7-8	243			8-9-3	337		
				9-1-5	339		
				7-2-6	346		

Tap Combination	Cycle Length	Tap Combination	Cycle Length	Tap Combination	Cycle Length	Tap Combination	Cycle Length
Group 9 400 to 449		Group 10 450 to 499		Group 11 500 to 549		Group 12 550 to 599	
9-4-3	400	8-9-1	450	9-1-10	510	8-9-10	553
8-7-2	400	6-5-7	460	10-8-9	517	10-4-2	555
8-5-4	402	7-3-9	469	9-6-10	521	7-8-5	561
8-5-10	443	6-3-10	481	8-2-3	527	6-3-4	561
7-4-1	445	8-3-5	483	7-10-8	528	9-2-1	573
8-9-6	445	6-8-3	483	8-10-2	549	7-9-8	576
6-9-8	445	7-6-9	495			9-8-1	581
10-7-4	446	8-1-4	496			10-3-2	581
		7-10-3	496			10-8-2	582
		6-8-5	499			6-8-9	586
		6-3-5	499			9-8-6	586
						10-6-8	599
Group 13 600 to 649		Group 14 650 to 699		Group 15 700 to 749		Group 16 750 to 799	
10-2-5	606	8-1-7	687	6-8-2	711	8-1-6	759
8-9-4	622	7-1-8	687	9-10-7	718	6-1-8	759
6-10-4	624			10-2-7	719	10-2-4	769
9-5-8	646			7-4-6	725	9-7-8	773
				6-10-2	733	9-4-2	776
				8-4-7	739	9-7-2	776
						7-6-4	778
						7-5-4	778
						9-5-7	783
						8-5-9	795
						10-3-5	796
						6-10-7	799

Table VI-20 (Cont'd).

Group 17 800 to 849		Group 18 850 to 899		Group 19 900 to 949		Group 20 950 to 999	
Tap Combination	Cycle Length	Tap Combination	Cycle Length	Tap Combination	Cycle Length	Tap Combination	Cycle Length
9-1-3	819	7-1-3	860	7-6-2	918	10-6-3	950
9-7-10	825	9-1-6	868	9-5-4	919	9-10-8	950
6-2-7	830	6-7-3	871	8-6-10	922	8-5-1	950
6-4-7	831	8-4-5	872	8-6-1	940	9-3-5	955
9-10-1	839	6-5-2	875	10-5-3	941	8-6-4	959
		9-6-5	876	9-3-10	941	7-5-3	959
		7-6-1	880	8-10-1	942	8-6-7	966
		10-5-4	881	10-1-3	943	9-5-2	968
		9-10-6	884	8-10-9	949	9-6-2	969
		8-7-4	893			9-1-4	969
		7-4-3	893			7-10-2	969
		10-5-7	897			6-1-9	973
						8-6-9	980
						9-6-8	980
						9-2-6	983
						6-2-9	983

Group 21 1000 to 1049		Group 22 1050 to 1099		Group 23 1100 to 1149		Group 24 1150 to 1199	
Tap Combination	Cycle Length	Tap Combination	Cycle Length	Tap Combination	Cycle Length	Tap Combination	Cycle Length
8-4-6	1036	6-4-2	1059	6-5-9	1107	7-8-10	1150
7-5-2	1040	9-7-5	1060	9-5-6	1107	10-8-7	1151
9-6-4	1041	6-8-10	1067	8-10-6	1112	8-10-5	1151
		10-8-6	1067	6-10-8	1112	6-5-8	1153
		10-7-3	1084	6-7-1	1119	8-5-6	1153
		8-4-1	1084	10-4-5	1120	8-5-7	1157
		6-9-3	1094	9-3-8	1121	7-5-8	1157
				8-3-9	1121	8-9-7	1161
				7-8-1	1123	7-8-9	1161
				6-8-1	1136	9-8-7	1162
				9-4-5	1138	7-1-6	1176
				6-7-2	1138	6-1-7	1176
				7-5-10	1141	8-10-3	1179
						8-1-3	1179

Tap Combination	Cycle Length	Tap Combination	Cycle Length	Tap Combination	Cycle Length	Tap Combination	Cycle Length
Group 25 1200 to 1249		Group 26 1250 to 1299		Group 27 1300 to 1349		Group 28 1350 to 1399	
10-3-6	1210	10-2-8	1258	6-4-9	1317	7-4-5	1364
7-9-2	1222	9-8-10	1263	9-4-6	1317	6-7-4	1364
8-1-2	1228	10-5-9	1279	7-10-1	1347	7-3-5	1368
7-10-6	1228	9-5-10	1279			10-9-1	1373
9-10-3	1229	10-6-2	1286			10-2-1	1373
8-6-3	1235	8-4-10	1292			7-5-6	1379
8-5-3	1235	9-3-7	1296			8-2-7	1391
						7-2-8	1391
Group 29 1400 to 1449		Group 30 1450 to 1499		Group 31 1500 to 1549		Group 32 1550 to 1599	
7-6-3	1401	6-1-2	1450	7-3-10	1503	6-3-8	1552
8-2-6	1403	10-5-2	1451	10-3-7	1503	10-9-8	1554
6-2-8	1403	10-7-2	1471	9-2-7	1506	9-5-1	1563
8-7-3	1433	8-7-5	1477	7-2-9	1506	10-2-3	1575
8-4-3	1433	6-4-3	1477	10-5-6	1517	7-6-5	1587
6-2-1	1444	10-6-9	1487	8-2-1	1538	6-5-4	1587
10-9-5	1445			10-9-3	1539	6-9-7	1592
				10-8-1	1541	10-2-6	1594
				10-3-1	1541	10-9-6	1595
						6-9-10	1595

Table VI-20 (Cont'd).

Tap Comb. (Group 33, 1600–1649)	Cycle Length	Tap Comb. (Group 34, 1650–1699)	Cycle Length	Tap Comb. (Group 35, 1700–1749)	Cycle Length	Tap Comb. (Group 36, 1750–1799)	Cycle Length	Tap Comb. (Group 37, 1800–1849)	Cycle Length	Tap Comb. (Group 38, 1850–1899)	Cycle Length	Tap Comb. (Group 39, 1900–1949)	Cycle Length	Tap Comb. (Group 40, 1950–1999)	Cycle Length	Tap Comb. (Group 41, 2000–2049)	Cycle Length
8-9-5	1606	6-10-5	1663	6-9-4	1704	7-3-2	1750	6 4 10	1802	9-2-5	1850	8-7-10	1904	7-9-4	1951	7-1-2	2013
6-2-3	1606	6-1-5	1663	6-2-10	1747	9-8-4	1751	10-4-6	1802	6-9-2	1850	10-7-8	1905	7-2-4	1951	9-10-4	2014
7-8-6	1623	10-1-5	1682			9-4-10	1766	9-6-1	1811	8-3-4	1857	7-10-4	1905	9-3-1	1958	6-7-5	2028
6-8-7	1623	10-1-5	1683			10-4-9	1766	7-9-5	1838	7-8-3	1857	7-1-4	1905	10-6-1	1966	6-4-5	2028
8-1-5	1627	9-2-3	1683			6-9-5	1770	6-2-4	1844	7-3-8	1866	7-10-5	1912	10-5-1	1966	10-7-1	2040
6-10-3	1627	8-9-2	1683			6-2-5	1770	9-3-6	1844	8-3-7	1866	6-1-4	1912	7-6-8	1966	10-4-1	2040
10-4-7	1630	9-7-1	1688			9-1-2	1778	6-3-9	1844	8-3-1	1888	7-4-2	1915			7-2-1	2043
9-8-10	1630					9-10-2	1778	7-6-10	1847	10-8-3	1889	9-7-4	1916			10 9-4	2044
7-8-4	1631					7-5-1	1779	10-6-7	1847	10-3-8	1889	8-2-9	1918			9-6-3	2044
9-1-7	1633					10-6-4	1780			8-3-10	1889	6-3-1	1928			8-5-2	2044
7-1-9	1633									9-4-1	1892	10-8-5	1929			7-9 1	2046
9-3-4	1641									6-1-10	1894						
7-8-2	1641									6-3-2	1897						
8-2-10	1646									9-8-5	1898						

Average cycle length = 1010.72

$2^{n-1} = 1024$

Chapter VII

CYCLES FROM NONLINEAR SHIFT REGISTERS

1. INTRODUCTION

Nonlinear shift registers are a convenient source of sequences of pseudorandom digits for numerous applications, including efficient communications, Monte Carlo statistical techniques, correlation guidance systems, and range radar pulse patterns. The results of initial theoretical and experimental investigations of shift register sequences were presented in Chapter VI.

From an *n*-stage shift register with feedback, it is possible to obtain one or more sequences, or *cycles*. The sum of the lengths of all the cyclically distinct cycles is 2^n. If more than one cycle is possible from a given shift register, the initial state of the register determines the specific cycle. For many applications, it is desirable to use cycles which closely approach the maximum length of 2^n. If the initial state of the shift register is chosen arbitrarily, the resulting cycle has an expected length of one-half the maximum. In fact, all lengths from 1 to 2^n have approximately equal probability for such a cycle. On the other hand, if care is exercised to assure that the longest cycle is being used for the given shift register, the expected length (averaged over possible shift registers) is about five-eighths of the maximum.

The cycle structure from certain highly simplified shift registers has been fully explored. Facts about the occurrence of cycles of specific lengths have been compiled. A statistical model of cycle distribution has been developed, and this model has been compared with both theoretical and experimental results for the shift register sequences. Also, a method is described for obtaining cycles of any preassigned length.

2. THE DETERMINISTIC APPROACH

The nonlinear shift register (Fig. VII-1) which mechanizes the recurrence relation

$$a_k = f(a_{k-1}, a_{k-2}, \ldots, a_{k-n+1}) \oplus a_{k-n} , \tag{1}$$

169

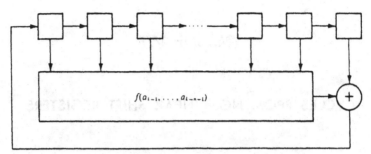

Fig. VII-1. The nonlinear shift register which mechanizes the recurrence relation Eq. (1).

where f is an arbitrary Boolean function of $n-1$ binary variables,
permutes the vectors of binary n-dimensional space V_n, if the vector
$(a_{k-1}, a_{k-2}, \ldots, a_{k-n})$ is viewed as having the vector $(a_k, a_{k-1}, \ldots, a_{k-n+1})$
as its successor. That is, the relation Eq. (1) assigns to each vector
of V_n a unique successor, and thereby decomposes V_n into cycles of
consecutive vectors. These facts were developed in Chapter VI, where
it was further proved that for $n > 2$, the number of cycles into which
V_n is decomposed is even or odd as the number of 1's in the truth
table of f is even or odd.

   As a simple example, with $n = 4$, consider the relation

$$a_k = a_{k-1}a_{k-2} \oplus a_{k-4}. \tag{2}$$

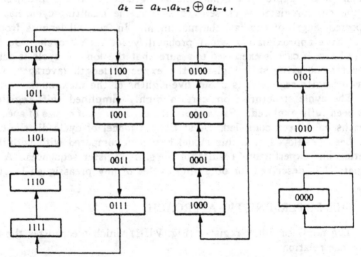

Fig. VII-2. The four possible cycles from the shift register which
mechanizes the recurrence relation Eq. (2).

Table VII-1. Truth table for the function
$f(a_{k-1}, a_{k-2}, a_{k-3}) = a_{k-1}a_{k-2}$.

a_{k-1}	a_{k-2}	a_{k-3}	f
0	0	0	0
0	0	1	0
0	1	0	0
0	1	1	0
1	0	0	0
1	0	1	0
1	1	0	1
1	1	1	1

The sixteen vectors of V_4 form into cycles as shown in Figure VII-2. The number of cycles for this shift register is even, because the truth table for $f(a_{k-1}, a_{k-2}, a_{k-3})$ in Eq. (2) contains an even number of 1's, as shown in Table VII-1.

It should be noted that whenever one or more of the variables in f fail to occur explicitly, the number of 1's in the truth table, and consequently the number of cycles for V_n, is necessarily even. This is also the case if one or more of the variables of f occur only linearly, and under certain other circumstances (see Chapter VI).

The cycle structure obtained from certain simplified shift registers will be discussed, as will the criteria for the existence of short cycles, which brings up to date the treatment given in Chapter VI. In addition, one outstanding conjecture concerning cycle structure will be presented.

2.1. *Cycles from simple registers*

The following four degenerate recursion formulas will be considered; it is incidentally true that all four of the formulas are linear:

$$a_k = a_{k-n} \tag{3}$$

$$a_k = 1 \oplus a_{k-n} \tag{4}$$

$$a_k = a_{k-1} \oplus a_{k-2} \oplus \cdots \oplus a_{k-n} \tag{5}$$

$$a_k = 1 \oplus a_{k-1} \oplus a_{k-2} \oplus \cdots \oplus a_{k-n}. \tag{6}$$

The corresponding shift registers are termed the *pure cycling register* (PCR), the *complemented cycling register* (CCR), the *pure summing register* (PSR), and the *complemented summing register* (CSR), respectively. The decompositions of V_4 into cycles for each of these cases are shown in Table VII-2, where the cycle notation $(ab...q)$ indicates that the sequence obtained is $ab...qab...qab...q...$, with periodic repetition.

As shown in Chapter VI, the number of cycles obtained from the PCR is

$$Z(n) = \frac{1}{n} \sum_{d|n} \phi(d)2^{n/d} , \tag{7}$$

where $\phi(d)$ is Euler's ϕ-function, and the summation is over all divisors d of n. It was further shown that $Z(n)$ is *even* for all $n \neq 2$.

Table VII-2. The cycle decompositions of V_4 induced by each of the four simple shift registers.

PCR	CCR	PSR	CSR
(0) (1)			
(01)	(00001111)	(0)	(1)
(0001)		(11110)	(00001)
(0011)	(01011010)	(11000)	(00111)
(0111)		(10100)	(01011)

A similar derivation for the CCR yields a number of cycles $Z^*(n)$, where

$$Z^*(n) = \tfrac{1}{2}Z(n) - \frac{1}{2n} \sum_{2d|n} \phi(2d)2^{n/2d} . \tag{8}$$

Here the new summation is in effect restricted to the *even* divisors $2d$ of n. Note that for odd n, $Z^*(n) = \tfrac{1}{2}Z(n)$. If $S(n)$ and $S^*(n)$ are the enumerators for the PSR and CSR cases, respectively, it is easily shown that

Table VII-3. The cycle enumerator functions for four simple shift registers.

n	PCR	CCR	PSR	CSR
	$Z(n)$	$Z^*(n)$	$S(n)$	$S^*(n)$
1	2	1	2	1
2	3	1	2	2
3	4	2	4	2
4	6	2	4	4
5	8	4	8	6
6	14	6	10	10
7	20	10	20	16
8	36	16	30	30
9	60	30	56	52
10	108	52	94	94

$$S(n) + S^*(n) = Z(n + 1), \tag{9}$$

because the cycles of the PSR and the CSR combined are identical with the cycles of the PCR for register length $n + 1$. It is further true that

$$S(n) = Z(n + 1) - Z^*(n + 1). \tag{10}$$

The explicit expressions for $Z^*(n)$, $S(n)$, and $S^*(n)$ are thus:

$$Z^*(n) = S^*(n - 1) = \frac{1}{2n} \sum_{\text{odd } d \mid n} \phi(d) 2^{n/d}, \tag{11}$$

and

$$S(n - 1) = \frac{1}{2n} \sum_{d \mid n} \phi(d) 2^{n/d} + \frac{1}{2n} \sum_{\text{even } d \mid n} \phi(d) 2^{n/d}. \tag{12}$$

These four functions are tabulated, for $n \leq 10$, in Table VII-3.

2.2. Criteria for short cycles

In terms of the recurrence relation Eq. (1), the cycle *all 0's* occurs in a shift register if and only if $f(0, 0, \ldots, 0) = 0$. Similarly, the cycle *all 1's* occurs if and only if $f(1, 1, \ldots, 1) = 0$. These criteria appear in Chapter VI, where it is further stated that the congruence

$$f(x_1, x_2, \ldots, x_{n-1}) = f(x_1', x_2', \ldots, x_{n-1}') \equiv n \pmod{2} \tag{13}$$

determines whether or not $01010101\cdots$ is a cycle of the shift register. Actually, Eq. (13) need only hold for $(x_1, x_2, x_3, x_4, \ldots) = (0, 1, 0, 1, \ldots)$. However, if $f(x_1, x_2, \ldots, x_{n-1}) - f(x_1', x_2', \ldots, x_{n-1}')$ is identically a constant, as it is for any linear function f and for other examples given in Chapter VI, the criterion Eq. (13) can be used to determine whole classes of recursion relations for which $010101\ldots$ is *always* a cycle, or *never* a cycle, irrespective of tap positions.

It is interesting to ask what restrictions exist on the capability of an n-stage shift register to have cycles of length n. We regard $0000\ldots$ as having period 1, rather than n, and in general no subperiods of n will be allowed. A precise condition for the existence of an n-cycle is that some n-vector pass through the shift register unmodified, as through the pure cycling register. That is, $f(x_1, x_2, \ldots) \oplus w = w$ must hold for a cyclic succession of vectors $(a, b, \ldots)$, $(b, c, \ldots)$, etc. That is,

$$f(x_1, x_2, \ldots) \equiv 0 \tag{14}$$

must hold for these vectors. Moreover, to avoid unit periodicity, at least one 1 and at least one 0 must occur in the vector $(a, b, \ldots)$. In fact, it is a consequence of "no subperiod" that a 1 and a 0 exist at any specified distance apart, where positions and distances are regarded modulo n. This is in itself enough information to rule out cycles of length n for many cases.

Example. Consider $f(x, y, z) = y \oplus z \oplus xy \oplus xz \oplus yz$, the Logic 1 of Chapter VI. If there is a cycle of length n, $f(x, y, z) \equiv 0$ must hold for those vectors which represent the n-cycle. But if there is no sub-period for one or more vectors of the cycle, we will have $x = 0$ and $y = 1$, no matter where the two taps x and y are placed. In such a case, $f(x, y, z) = 1 \oplus z \oplus z = 1$, which precludes the existence of an n-cycle.   '

A rather obvious theorem asserts that the total number of cycles of length $\leq n$ (i.e., the total number of "short cycles") cannot exceed $Z(n)$ for any n-stage shift register. In particular, $Z(n)$ is the total number of cycles for all lengths d which divide n. Now, if there is a cycle present of length less than n, and whose length does not divide n, it excludes the occurrence of at least one cycle whose length does divide n, so that the substitution of cycles of lengths not dividing n will never increase the total number of short cycles. For example, if there is a cycle $(abc\cdots m)$ of length $n - 1$ present, the cycle $(abc\cdots ma)$ of length n cannot occur, because the existence of the $n - 1$ cycle implies that after $abc\cdots ma$ occurs, it is followed by $bc\cdots m$ rather than by $abc\cdots m$. It can further be shown that distinct cycles of length not dividing n exclude distinct cycles of length dividing n, which completes the proof.

A similar theorem, propounded by W. Kautz and B. Elspas of the Stanford Research Institute, asserts that the total number of cycles of length less than n (strict inequality) cannot exceed

$$\frac{1}{n} \sum_{d|n} \mu(d) 2^{n/d} , \tag{15}$$

which is also the number of n-length cycles obtained from the pure cycling register. Here $\mu(d)$ is the Möbius function.

2.3. *An outstanding conjecture*

The following conjecture seems highly plausible, but no promising method of proof has yet been discovered. It is likely that the proof or disproof of this statement would contribute a deeper insight into certain aspects of shift register theory.

Conjecture A. The maximum number of cycles into which V_n can be decomposed by an n-stage shift register is equal to $Z(n)$.

Since the PCR attains this number, the problem is to show that no larger number is possible. From Table VII-3, it is seen that the PSR attains the same number of cycles when $n = 1, 3$, and 5 as the PCR, but with a different cycle decomposition. Hence, if $Z(n)$ is indeed the maximum, it is still not a uniquely attained maximum in all cases. Conjecture A is basically a statement concerning paths in the Good-de Bruijn diagram.

3. THE STATISTICAL APPROACH

In the states generated by a shift register, there is clearly a strong statistical dependence between any vector and its immediate successor. However, it has proven interesting to explore the consequences of the mathematical model wherein successors of vectors are chosen at random, and to compare this situation with the experimentally observed data. In this section, the mathematical model and its properties are described. Extensive experimental data are available for comparison, and it is found that the model is very reliable insofar as the number of cycles and the expected lengths of long cycles are concerned, but inaccurate concerning the repetition of cycle lengths and the exclusion of short cycles of certain lengths. Also treated are certain related statistical models which arise in the study of nonlinear shift register sequences.

3.1. *The mathematical model*

A random permutation on a set of N objects can be characterized as follows:

Let the N objects of the set be arbitrarily numbered $1, 2, 3, \ldots, N$. Let these objects be sampled randomly, without replacement, until object 1 is obtained. Then the sampled objects $a_1, a_2, \ldots, a_{m_1-1}, 1$, are said to form the *first cycle*, of length m_1, where each object has a successor under the convention that 1 is followed by a_1. Let x be the lowest-numbered object (if any) not contained in the first cycle. From the set of $N - m_1$ objects which remain after the first cycle has been extracted, the objects are sampled without replacement until x is obtained. The objects $b_1, b_2, \ldots, b_{m_2-1}, x$, are said to form the *second cycle*. This process is continued until all N objects have been placed on cycles.

The sampling process just described will be termed M for brevity. When the objects being sampled are the $2^n = N$ binary vectors of length n, the process M serves as a simplified model for the decomposition of binary n-space into cycles by a nonlinear shift register which satisfies the cycles condition of Eq. (1). In the ensuing discussion, properties of the model M will be deduced, and these will then be compared with experimental data on cycles obtained from actual shift register operation.

3.2. *Simple consequences of the model*

Theorem 1. The distribution of the lengths m_1 of the first cycles obtained from N objects by the sampling process M is *flat*. That is, all values of m_1 from 1 to N are equally likely, each having probability equal to $1/N$.

Proof. The probability of selecting 1 on the first try is clearly $1/N$, so that $\mathrm{pr}\,(m_1 = 1) = 1/N$. Using complete induction, suppose it is established that $\mathrm{pr}\,(m_1 = 1) = \mathrm{pr}\,(m_1 = 2) = \cdots = \mathrm{pr}\,(m_1 = k) = 1/N$, where it must then be shown that $\mathrm{pr}\,(m_1 = k + 1) = 1/N$. The probability $\mathrm{pr}\,(m_1 = k + 1)$ that 1 is selected on the $(k + 1)$st trial is the product of two probabilities: $p_1 p_2$, where p_1 is the probability that the selection process has not already terminated before the $(k + 1)$st stage, and p_2 is the probability that if the process has reached the $(k + 1)$st stage, then 1 will be the next selection. By the complete induction hypothesis, $p_1 = 1 - k/N$; and because 1 is in a collection of $N - k$ equally likely choices at the time of the $(k + 1)$st selection, $p_2 = 1/(N - k)$. Thus

$$\mathrm{pr}\,(m_1 = k + 1) = p_1 p_2 = \left(1 - \frac{k}{N}\right)\frac{1}{N - k} = \frac{1}{N}\,,$$

completing the inductive proof.

Corollary 1. The expected length $\bar{m}_1$ of the first cycle, and the standard deviation σ about this mean, are given by

$$\bar{m}_1 = \tfrac{1}{2}(N + 1)$$
$$\sigma^2 = (N^2 - 1)/12\,. \tag{16}$$

Proof. These formulas for the centroid and moment of inertia of a flat distribution are derived in every elementary mechanics, physics, and probability text.

Theorem 2. The expected number $\bar{C}$ of cycles obtained from N objects by the sampling process M is

$$\bar{C} = \sum_{k=1}^{N}\frac{1}{k} = \ln N + \gamma + o(1)\,, \tag{17}$$

where γ is Euler's constant, $0.5771\ldots$.

Proof. When $N = 1$, the expected number of cycles is $\sum_{k=1}^{1} 1/k = 1$. The theorem will therefore follow by induction if it can be shown that $C_N = C_{N-1} + 1/N$, where C_k is the expected number of cycles for the case $N = k$. But clearly,

$$C_N = 1 + \frac{1}{N}\sum_{m=1}^{N} C_{N-m} \tag{18}$$

since if the first cycle, which has length m for some $1 \le m \le N$, is removed, the expected number of remaining cycles is C_{N-m}. Weighting each C_{N-m} by the corresponding probability, which is $1/N$ for all values of m by Theorem 1, and adding, the result is the expected number of cycles after the first one has been removed. Adding back 1 for the contribution of the first cycle to the total then yields Eq. (18).

By elementary manipulation from Eq. (18), using $C_0 = 0$,

$$NC_N = N + \sum_{i=1}^{N-1} C_i$$

$$(N-1)C_{N-1} = N - 1 + \sum_{i=1}^{N-2} C_i$$

$$NC_{N-1} = N - 1 + \sum_{i=1}^{N-1} C_i$$

$$N(C_N - C_{N-1}) = 1$$

$$C_N = C_{N-1} + \frac{1}{N},$$

which completes the proof.

Another proof of Theorem 2, as well as a proof of Theorem 3, is given in [5] (pp. 205-206).

Theorem 3. The distribution of cycle lengths, shown in Theorem 2 to have a mean of $\sum_{1}^{N} 1/k$, has a variance about this mean of

$$\sigma^2 = \sum_{k=1}^{N} \frac{N-k}{(N-k-1)^2} = \ln N + \gamma + \frac{\pi^2}{6} + o(1). \tag{19}$$

In his recent book on combinatorial analysis [32], Riordan devotes an entire chapter to the distribution of cycle lengths from random permutations. In particular, proofs of Theorems 2 and 3 are presented. Many of Riordan's results are of no particular interest here (specifically those involving the presence or absence of cycles of length 1, since in such respects the statistical model bears no resemblance to the actual behavior of shift registers); however, one of the theorems is well worth inclusion:

Theorem 4. Among the $N!$ decompositions of N objects into cycles, the number of cases with exactly k cycles is $c(N, k)$, where the numbers $c(N, k)$ are the Stirling numbers of the first kind, generated by $t(t + 1)(t + 2)\cdots(t + N - 1) = \sum_{k=0}^{N} c(N, k)t^k$.

The Stirling numbers for N not exceeding 8 are presented in Table VII-4. The probability of exactly k cycles is of course $c(N, k)/N!$

From Table VII-4, it is seen that the probability of everything lying on a single cycle is $c(N, 1)/N! = (N - 1)!/N! = 1 N$, which is also the special case of Theorem 1 with $m_1 = 1$. The probability of two cycles when $N = 8$ is 13068/8!, etc. Note that Theorem 4 describes the *distribution* for which Theorems 2 and 3 specify the mean and the variance.

Table VII-4.   Stirling numbers of the first kind, $c(N, k)$.

N	k							
	1	2	3	4	5	6	7	8
1	1							
2	1	1						
3	2	3	1					
4	6	11	6	1				
5	24	50	35	10	1			
6	120	274	225	85	15	1		
7	720	1764	1624	735	175	21	1	
8	5040	13068	13132	6769	1960	322	28	1

3.3.   *The expected length of the longest cycle*

For present purposes, one of the most significant questions is the expected length of the longest cycle. It is surprising that this problem has virtually escaped attention in the literature (including Riordan's book). The only treatment at all appears to be in Goncharov's Russian article [16], where a generating function for the problem is described, but no numerical answers are obtained.

Throughout this section, it is assumed than N objects are distributed into cycles by the selection process M.

> *Theorem 5.* Let L_N be the expected length of the longest cycle, which then has *relative expected length* L_N/N. In the limit as $N \to \infty$, the probability that the first cycle is longest equals the relative expected length L_N/N of the longest cycle.

Proof. The choice of which element is "first" is in fact an arbitrary one, and therefore the chance of the first element being on a cycle of length T is T/N, for all T. The specific case of being on the longest cycle, which has expected length L_N, appears to lead to Theorem 5. The only difficulty involves the meaning of "longest cycle", since this may not be unique. If we consider the first cycle to be "longest" whenever it is *tied* for longest, then the probability that it is longest *exceeds* the ratio L_N/N, although in the limit as $N \to \infty$, equality occurs.

Theorem 6. The constant λ, given by $\lim\limits_{N \to \infty} (L_N/N) = \lambda$, exists. Moreover, $0.59523 < \lambda < 0.7071$.

Proof. Let $b_N = L_N/(N + 1)$. Then

$$b_{N-1} \geq b_N , \tag{20}$$

because when an $(N + 1)$st number is added to the original set of N elements, it has a probability $L_N/(N + 1)$ of lying on what is already the longest cycle, increasing the expectation from L_N for populations of size N to at least $L_N + L_N/(N + 1)$ for populations of size $N + 1$. (Since it is possible for two different cycles to be tied for first place, it is actually true that the extra member contributes *more* to the expectation than $L_N/(N + 1)$.) Rewriting $L_{N+1} \geq L_N + L_N/(N + 1)$ as $L_{N+1} \geq L_N(N + 2)/(N + 1)$, and this in turn as $L_{N+1}/(N + 2) \geq L_N/(N + 1)$, the inequality (Eq. 20) for the b's results.

By a theorem of Weierstrass, since the sequence $\{b_N\}$ is a monotone nondecreasing sequence, bounded from above by 1, $\lim\limits_{N \to \infty} b_N = \lambda$ exists. Moreover, since $\lim (N + 1) N = 1$, we obtain

$$\lim \frac{L_N}{N} = \lim \frac{L_N}{N + 1} \frac{N + 1}{N} = \lim b_N = \lambda$$

where all limits are as $N \to \infty$.

It can in fact be seen that $\lambda \leq 0.7071$, since from Theorem 5, $\lambda_N =$ probability that the longest cycle is first = expected relative length L_N/N of the longest cycle; while from Corollary 1, the expected relative length F_N/N of the *first* cycle is only 0.5. Clearly

$$F_N \geq L_N \lambda_N , \tag{21}$$

because the expected length of the first cycle is the sum of the expected length of each cycle, times the probability that the cycle in question comes first; and $L_N \lambda_N$ does not exceed the particular summand corresponding to the longest cycle.

From Eq. (21), $\lambda_N^2 \leq 0.5$, whereby $\lambda \leq 2^{-1/2} = 0.7071\dots$. In the other direction, $b_{10} = 23{,}759{,}791/11! = 0.59523286$, and $\lambda \geq b_N$ for all N.

Note: In contrast to the nondecreasing sequence b_N, it appears that λ_N is a nonincreasing sequence, which, if true, could be used as an alternative proof for the existence of λ.

The exact value of λ is not known in closed form, nor in terms of familiar constants, but the following results narrow the range considerably. (See also Section 4, pp. 189–192.)

Theorem 7. Among the $N!$ distinguishable cycle decompositions of N objects, there are exactly $N!/p$ cycles of length p, for $1 \leq p \leq N$.

Proof. There are $\binom{N}{p}$ ways to choose the p objects to go on the cycle; there are $(p - 1)!$ cyclically distinct ways to arrange them; and there are $(N - p)!$ ways of handling the remaining $N - p$ objects. Moreover, $\binom{N}{p}(p - 1)! \, (N - p)! = N!/p$.

Corollary 7.1. The expected number of cycles per decomposition is $\sum\limits_{p=1}^{N} 1 \cdot p$. (This is merely a new approach to Theorem 2.)

Proof. The expected number of cycles is $(1.N!) \sum\limits_{p=1}^{N} N! \, p = \sum\limits_{p=1}^{N} 1 \, p$.

Corollary 7.2. The expected number of p-cycles per decomposition is $1/p$.

Proof. There are $N!/p$ p-cycles distributed among $N!$ cases.

Definition. Among the $N!$ decompositions, let $a_p N!$ be the number of cases where the longest cycle has length p.

The quantity a_p may also be regarded as the probability that for a random permutation the longest cycle has length p. Thus $\sum\limits_{p=1}^{N} a_p = 1$; and by Theorem 7, $0 < a_p \leq 1 \, p$ for $1 \leq p \leq N$. Clearly $a_p = 1 \, p$ for $\frac{1}{2}N < p \leq N$, because a cycle which contains more than half the possible elements is surely longest. Also, $a_p < 1'p$ for $1 \leq p \leq \frac{1}{2}N$; and $a_1 = 1/N!$.

Theorem 8. $\lambda \leq 1 - 1/e = 0.632120\ldots$.

Proof. The expected length of the longest cycle is $L_N = \sum\limits_{p=1}^{N} p a_p$. Since $\sum\limits_{p=1}^{N} a_p = 1$, and $a_p = 1/p$ for $p = 1, 2, \ldots, N$, it follows that $L_N \leq \sum\limits_{p=N-x+1}^{N} p(1/p) = x$, where x is the smallest integer such that $\sum\limits_{p=N-x-1}^{N} 1 \, p \geq 1$. That is, $\log_e [N/(N - x)] = 1 + o(1)$, and $N/(N - x) \sim e$, as $N \to \infty$. This leads to $x/N \sim 1 - 1/e$, $L_N/N \leq x/N \sim 1 - 1/e$, and $\lambda = \lim\limits_{N \to \infty} L_N \, N \leq 1 - 1'e$.

Theorem 9. $\lambda \geq \dfrac{2}{3} + \dfrac{1}{3} \log_e \dfrac{27}{32} = 0.61003\ldots$.

Proof. For $\frac{1}{2}N < p \leq N$, it is clear that

$$\frac{1}{N} \sum_{p > 1/2 \, N}^{N} p a_p \sim \frac{1}{2},$$

since $a_p = 1/p$. For $\frac{1}{3}N < p \leq \frac{1}{2}N$, it is found that

$$a_p = \frac{1}{p}\left(1 - \frac{1}{2}p - \sum_{q=p-1}^{N-p} \frac{1}{q}\right),$$

based on the fact that the probability of a cycle of length $q > p$ is $\sum_{q=p-1}^{N-p} 1/pq$, and the fact that with probability $1/p^2$ a given cycle of length p is tied for longest with another p-cycle. The contribution to the expected length of the longest cycle is therefore

$$\frac{1}{N} \sum_{(1/3)N < p \leq (1/2)N} p a_p = \frac{1}{N} \sum_{(1/3)N < p \leq (1/2)N} \left(1 - \frac{1}{2p} - \sum_{p+1}^{N-p} \frac{1}{q}\right)$$

$$= \frac{1}{N}\left(\frac{N}{6} + \sum_{(1/3)N < p \leq (1/3)N} \log_e \frac{N-p}{p}\right) + o(1),$$

which equals

$$\frac{1}{6} + \frac{1}{N}\left[-\log_e\left(\frac{2N}{3e}\right)^{(2/3)N} + 2\log_e\left(\frac{N}{2e}\right)^{(1/2)N}\right.$$

$$\left. - \log_e\left(\frac{N}{3e}\right)^{(1/3)N}\right] + o(1) = \frac{1}{6} + \frac{1}{3}\log_e \frac{27}{32} + o(1).$$

Thus,

$$\frac{1}{N} \sum_{1 \leq p \leq N} p a_p \geqq \frac{1}{N} \sum_{(1/3)N < p \leq N} p a_p \sim \frac{1}{2} + \left(\frac{1}{6} + \frac{1}{3}\log_e \frac{27}{32}\right),$$

so that

$$\lambda \geq \frac{2}{3} + \frac{1}{3}\log_e \frac{27}{32}.$$

Theorem 10.

$$\lambda \leq 1 + \frac{1}{3}\log_e \frac{27}{32} - \frac{1}{e\log_e(32e/27)} = 0.62891\ldots.$$

Proof. First,

$$\sum_{p > (1/2)N} a_p \sim \log_e 2.$$

Next,

$$\sum_{(1/3)N < p \leq (1/2)N} a_p \leq \sum_{y < p \leq (1/2)N} \frac{1}{p} = \log_e \frac{N}{2y} + o(1),$$

where y is determined from

$$\frac{1}{N} \sum_{y < p \leq (1/2)N} p \frac{1}{p} \sim \frac{1}{6} + \frac{1}{3}\log_e \frac{27}{32}.$$

That is,

$$\frac{1}{2} - \frac{y}{N} \sim \frac{1}{6} + \frac{1}{3}\log_e \frac{27}{32}.$$

and

$$\frac{y}{N} \sim \frac{1}{3} \log_e \frac{27e}{32} .$$

Thus

$$\sum_{(1/3)N < p \le (1/2)N} a_p \le -\log_e \left(\frac{2}{3} \log_e \frac{27e}{32} \right) + o(1) .$$

Hence,

$$\sum_{1 \le p \le (1/3)N} a_p \le 1 - \log_e 2 + \log_e \left(\frac{2}{3} \log_e \frac{27e}{32} \right) + o(1)$$

$$= \log_e \left(\frac{e}{3} \log_e \frac{27e}{32} \right) + o(1) .$$

Then,

$$\frac{1}{N} \sum_{1 \le p \le (1/3)N} p a_p \ge \frac{1}{N} \sum_{p=z}^{(1/3)N} p \frac{1}{p} = \frac{1}{3} - \frac{z}{N} ,$$

where z is chosen to satisfy

$$\sum_{p=z}^{(1/3)N} \frac{1}{p} = \log_e \left(\frac{e}{3} \log_e \frac{27e}{32} \right) + o(1) ,$$

which leads to

$$\log_e \frac{N}{3z} = \log_e \left(\frac{e}{3} \log_e \frac{27e}{32} \right) + o(1)$$

and

$$\frac{z}{N} \sim \frac{1}{e \log_e (27e/32)} .$$

Hence,

$$\lim_{N \to \infty} \frac{1}{N} \sum_{1 \le p \le (1/3)N} p a_p \ge \frac{1}{3} + \frac{1}{e \log_e (32e/27)} ,$$

and

$$\lambda = \lim_{N \to \infty} \frac{1}{N} \sum_{1 \le p \le N} p a_p \ge \left(\frac{2}{3} + \frac{1}{3} \log_e \frac{27}{32} \right) + \left(\frac{1}{3} + \frac{1}{e \log_e (32e/27)} \right)$$

$$= 1 + \frac{1}{3} \log_e \frac{27}{32} + \frac{1}{e \log_e (32e/27)} .$$

Note: Successively finer approximations of this sort are possible, but become progressively more difficult. Moreover, such an approach cannot hope to determine λ exactly in a finite number of iterations. A digital computer program was used to obtain $\lambda = 0.62432965...$, which is sufficient to rule out the possibility of λ equaling 5/8, or $1 - 1/e$, or $(5^{1/2} - 1)/2$, or $(\pi/8)^{1/2}$. Whether or not λ bears any algebraic relationship to well-known mathematical constants is not yet established. It seems reasonable to conjecture at this point that no such relationship

Table VII-5. The approximating sequences for the
expected longest cycle.

n	$n!L_n$	λ_n	b_n	A_n
1	1	1.0000000000	0.5000000000	0.7500000000
2	3	0.7500000000	0.5000000000	0.6250000000
3	13	0.7222222222	0.5416666666	0.6319444444
4	67	0.6979166666	0.5583333333	0.6281250000
5	411	0.6850000000	0.5708333333	0.6279166666
6	2,911	0.6738425925	0.5775793650	0.6257109788
7	23,563	0.6678854875	0.5843995535	0.6261425205
8	213,543	0.6620256696	0.5884672619	0.6252464657
9	2,149,927	0.6582913849	0.5924622464	0.6253768157
10	23,759,791	0.6547561452	0.5952328584	0.6249945018

exists. The sequences $\{\lambda_n\}$ and $\{b_n\}$, which approximate λ from above and below respectively, and their average A_n, are tabulated for $n \leq 10$ in Table VII-5. The geometric mean $G_n = (\lambda_n b_n)^{1/2}$ appears to converge even more rapidly than A_n. One may also speculate as to whether the sequence $\{e_n\}$ converges, where e_n is defined by $\lambda = L_n/(n + e_n)$. The simplest conjecture here is $e_n \to \frac{1}{2}$.

Several refinements in the description of the distribution of the cycle lengths are indicated in the following remarks, for large N.

1. In order of sampling, the first cycle obtained has an expected length of $N/2$. The expected length of the next cycle (as determined by the process M) is then $\frac{1}{2}(N - N/2) = N/4$; and in general, the kth cycle sampled has an expected length of $N/2^k$.

2. In order of length, the longest cycle has an expected length of λN. The second longest cycle has an expected length close to $\lambda(1 - \lambda)N$, and with progressively decreasing accuracy, the kth-longest cycle has an expected length of $\lambda(1 - \lambda)^{k-1}N$.

3. It follows from Theorem 1 that the probability of the first cycle having a relative length $F/N > a$ is exactly $1 - a$, for $0 \leq a \leq 1$. The corresponding result for the longest cycle is not quite so simple. It is contained in the following theorem, which is presented here without proof. (See, however, Sec. 4.)

Theorem 11. The probability of the longest cycle having a relative length $L/N > a$ is exactly $\log_e (1/a)$, for $\frac{1}{2} \leq a \leq 1$.

Corollary 11. The probability that the longest cycle exceeds $\frac{1}{2}N$ in length is $\log_e 2 = 0.69315$, compared to the probability 0.50000 that the first cycle exceeds $\frac{1}{2}N$ in length.

This is merely the special case of Theorem 11 with $a = \frac{1}{2}$.

3.4. *Analysis of the experimental results*

The recurrence relation

$$a_k = y \oplus z \oplus xy \oplus xz \oplus yz \oplus w , \tag{22}$$

where $w = a_{k-n}$, and x, y, z are any three distinct members of the set $a_{k-1}, a_{k-2}, \ldots, a_{k-n+1}$, is a special case of the general recurrence relation

$$a_k = f(a_{k-1}, a_{k-2}, \ldots, a_{k-n+1}) \oplus a_{k-n} , \tag{1}$$

taking $f = y \oplus z \oplus xy \oplus xz \oplus yz$. This function f, which has the truth table shown in Table VII-6, was extensively studied in Chapter VI with primary emphasis on the distribution of the lengths of the first cycles, which were obtained using fixed n and a large selection of tap combinations for x, y, and z. In the single case of $n = 7$, *all* cycles were found for each of the 30 distinct tap combinations. (These results appeared in Chapter VI, Section 4.5.) In order to obtain a reasonable measure of statistical significance, a similar program was subsequently undertaken for $n = 11$.

Table VII-6. Truth table for the function
$f(x, y, z) = y \oplus z \oplus xy \oplus xz \oplus yz$.

x	y	z	f
0	0	0	0
0	0	1	1
0	1	0	1
0	1	1	1
1	0	0	0
1	0	1	0
1	1	0	0
1	1	1	1

There are several simple properties of the cycle decomposition, depending on the truth table, most of which were pointed out in Chapter VI:

1. The total number of cycles from any given tap combination is even, because the number of 1's in the truth table for f is even.

2. $00\ldots0$ is always a cycle, while $11\ldots1$ is never a cycle. (These are the only possible cases of unit cycles.)

3. $101010\ldots$ is never a cycle. (Hence, there are never any cycles of length 2.)

4. No cycles of length n can occur, where n is the register length. (This was shown in Sec. 1.)

In the cases tested for $n = 11$, cycles of all lengths from 1 to about $5n$ occurred, with the exceptions of length 2 and length n as just noted. A well-known consequence [5], [32] of the statistical model M used in Sections 3.2 and 3.3 is that the expected number of unit-length cycles is one per decomposition. (See also Corollary 7.2.) In the experimental work, there was exactly one unit-length cycle per decomposition, corresponding to the *all 0's* vector. According to the model, the occurrence of cycles of length k decreases in proportion to $1/k$ (Corollary 7.2), and on this point the experimental results were in basic agreement. Good agreement with the model was also achieved for each of the phenomena listed in Table VII-7.

Table VII-7. Phenomena exhibiting good agreement between model prediction and experimental data.

Phenomenon	Model Statistics	Experimental Result
Expected length of first cycle	50.0 %	48.22%
Expected length of longest cycle	62.4 %	61.09%
Probability that longest cycle is first	62.4 %	60.00%
Expected number of cycles	7.20	7.83
Median number of cycles	7	7
Expected length, second-longest cycle	23.56%	21.85%
Probability that first cycle exceeds $N/2$	50.0 %	45.00%
Probability that longest cycle exceeds $N/2$	69.3 %	70.00%

The experimental values in Table VII-7 correspond to $n = 11$, with a sample size of 180, representing the 180 independent tap locations for x, y, and z. The poorest agreement between model and experiment in Table VII-7 occurs in connection with the flatness of the distribution of the first cycles. However, it is precisely this property which was observed to hold in Chapter VI when a significantly larger sample, containing several thousand distinct tap combinations, was used.

The actual data from the experiments discussed here appear in Section 7, following a description of the experimental techniques, in Section 6.

3.5. *Anomalies in the experimental results*

The experimental results revealed certain unsuspected phenomena, not derivable from the simplified statistical model considered thus far. Most striking of these phenomena involved the circumstances under which several short cycles of the same length would occur for a fixed tap combination. The maximum number of possible cycles of length k is

$$\frac{1}{k} \sum_{d|k} \mu(d) 2^{k/d} \tag{23}$$

which serves as an upper bound to the number of times k can appear as a cycle length for any fixed tap combination.

It is noteworthy that this upper bound was never actually attained in the experimental data. Thus, only one of the two possible unit cycles could occur: 00...0, but not 11...1. The only two-cycle case, 101010..., was excluded. (These two instances were mentioned previously.) Moreover, whereas the two three-cycles, 100100100... and 110110110..., appeared separately, they were never found together for the same tap combination. Similarly, the three four-cycles, the six five-cycles, etc., never occurred simultaneously. These are consequences of the specific recurrence relation Eq. (22), and not of the statistical model.

The most striking phenomenon uncovered by the experiments, however, was not the tendency of cycles of a given length to exclude others of that length, but precisely the opposite effect. The case of repeated cycle length occurred in twenty-three of the one hundred and eighty experiments. Moreover, for these twenty-three instances, the average number of repeated cycle lengths was 5.56, where a cycle length was considered repeated if the same number occurred two or more times. (Thus, for $x = 2$, $y = 4$, $z = 7$, the decomposition 1862, 54, 41, 18, 18, 18, 15, 7, 7, 4, 3, 1 is regarded as having five repeated cycle lengths.) The departure from the randomness model is evident to any poker player: although 87 per cent of the hands dealt do not contain so much as a single pair, the remaining 13 per cent of the hands average better than a full house! The circumstantial evidence indicates that someone is stacking the deck. Closer examination reveals that in eighteen out of the twenty-three cases where repeated lengths occur, the tap positions satisfy the equation

$$y + z = n, \tag{24}$$

and in only two cases where Eq. (24) is satisfied do repeated cycle lengths *fail* to occur. In three of the remaining five cases, a relation of the type $x + y = n$ or $x + z = n$ holds, although this bias is less extreme. The remaining two cases involve multiple repetition of the cycle length 22, which is twice the register length.

Although the manner in which the relation Eq. (24) causes cycle lengths to repeat has not been analyzed in detail, it is clear that basically $y + z = n$ allows certain cycles (e.g., $11001\ldots$) to be reversed (e.g., 10011) and still satisfy the recurrence relation. For the purpose of generating long sequences, the condition Eq. (24) is a good one to avoid, because of the many short cycles which occur in the decompositions.

Three other significant biases in the distribution of the cycle lengths were noted. One of these is the tendency for consecutive integers to occur as cycle lengths, even in those cases where repeated cycles do not occur. The second bias is the somewhat smaller tendency for cycle lengths differing by two to occur. The third anomaly, noticeable especially in cases containing repeated cycles, but in certain other cases as well, is the existence of large common factors for many of the cycle lengths which occur.

3.6. *Related statistical models*

When the function f in the recurrence relation Eq. (1) satisfies the constraint

$$f(a_{k-1}, a_{k-2}, \ldots, a_{k-n+1}) = f(a'_{k-1}, a'_{k-2}, \ldots, a'_{k-n+1}), \qquad (25)$$

where the prime denotes complementation, something interesting happens. Specifically, let the vector $v = (a_{k-1}, a_{k-2}, \ldots, a_{k-n})$ have as its successor the vector $v_1 = (a_k, a_{k-1}, \ldots, a_{k-n+1})$. Then the vector $v' = (a'_{k-1}, a'_{k-2}, \ldots, a'_{k-n})$ has as its successor the vector $v'_1 = (a'_k, a'_{k-1}, \ldots, a'_{k-n+1})$. That is, in view of Eq. (25), applying the recurrence relation Eq. (1) to v' yields a new term

$$\begin{aligned} b_k &= f(a'_{k-1}, a'_{k-2}, \ldots, a'_{k-n+1}) \oplus a'_{k-n} \\ &= f(a_{k-1}, a_{k-2}, \ldots, a_{k-n+1}) \oplus a_{k-n} \oplus 1 \\ &= a_k \oplus 1 = a'_k . \end{aligned}$$

A convenient formulation of this phenomenon is the statement that the successor operator commutes with the complementation operator.

The consequences of this fact are as follows. If v and v' lie on different cycles C and C', then C and C' have the same length and are exact complementary images of each other. On the other hand, if v and v' lie on the same cycle C'', then they are located 180° distant on C'', and C'' is even in length, with any connected half of C'' being the complement of the succeeding half.

Example. The function $f(x, y) = x \oplus y$ satisfies $f(x, y) = f(x', y')$. Thus, any three-tap linear logic $a_k = x \oplus y \oplus w$ (and more generally, any odd-tap linear logic) is an instance of the phenomenon under discussion. To be more specific, consider the recurrence relation

$$a_k = a_{k-1} \oplus a_{k-3} \oplus a_{k-4} . \qquad (26)$$

The corresponding cycle decomposition of V_4 is shown in Figure VII-3.

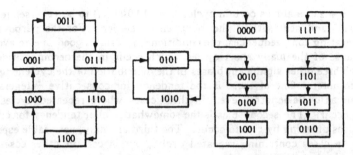

Fig. VII-3. The cycle decomposition of V_4 for the shift register
with the recurrence relation Eq. (26).

Among the six cycles there are two pairs of complementary cycles,
and two self-complementary cycles. Naturally, it is easy to give examples of nonlinear logics which also satisfy Eq. (25).

Instead of the model M for the distribution of $N = 2^n$ vectors, we
are led here to consider the model B for distributing $K = 2^{n-1}$ pairs of
vectors. We may think of K cards, each having a vector v on one
side, and the complementary vector v' on the reverse side. The process
B considers cards numbered from 1 to K on the front, and from $K + 1$
to $2K$ on the back, respectively. For any number x between 1 and
$2K$, we denote by x' that number $x \pm K$ which is also in the set, and
specify that x and x' are on opposite sides of the same card. The K
cards are placed in a hat and sampled randomly without replacement,
each sampled card being placed face up or face down with equal probability. The cards are sampled until the card containing the number
1 is drawn. If this card appears face up, the first cycle is $x_1, x_2, \ldots,$
$x_{m-1}, 1$. If it appears face down, the first cycle is defined to be $x_1, x_2,$
$\ldots, 1', x_1', x_2', \ldots, x_{m-1}', 1$. In the former case, $x_1', x_2', \ldots, x_{m-1}', 1'$ is
regarded as a *second* cycle.

The model B can be largely analyzed in terms of the model M,
which it actually becomes if complementation is ignored. Thus the
distribution of the first cycles for B involves a flat distribution for m,
$1 \le m \le K$ (from Theorem 1), followed by a decision with probability
0.5 either way as to whether m should be doubled or left alone. The
resulting distribution on 1 to $2K$ has two levels, namely $3/4K$ for all
lengths up to and including K, but only $1/4K$ for all lengths exceeding
K. This 3-to-1 behavior not merely agrees with the experimental results for such shift registers, but in fact was first observed experimentally. Another approach to this unusual result is that for cycle
lengths up to half of 2^n, the cycle can either be its own complement
or another cycle's complement; but as soon as it exceeds half of 2^n in

length, it is restricted to the case of being its own complement. The two-level distribution is thus a sum of two flat distributions, one on the interval 1 to K, and the other on the interval 1 to $2K$, with both summand distributions equally weighted (i.e., contributing equal areas). Since the basic properties of the statistical model B are derivable from the properties of the model M, an independent treatment is not warranted here.

Another variation of M is the following model $Q = Q_k$. There are N objects in the hat which are randomly sampled without replacement. However, any of k distinct objects will end the first cycle. It is required to find the probability distribution for the length of the first cycle. In the special case that $k = 1$, the model Q_1 reduces to the model M.

One application of the model Q was in the detection of equipment malfunction in the experimental work on cycles. The sequence generator was started with an initial vector v_0, and clock intervals were counted until v_0 reappeared. However, if the *and*-gate for the redetection of v_0 deteriorated and recognized one or more variants of v_0 as authentic (say k in all), the first cycle statistics become those of Q_k rather than those of M, with a noticeable bias in favor of the shorter cycles, even for $k = 2$.

The distribution of first cycles for Q_k is obviously the same as the distribution which solves the following problem, which recently appeared in *The American Mathematical Monthly* [14]:

"A quiz contestant selects a category containing N questions, k of which are too difficult for him. The questions are selected from the category at random, and the contestant continues answering until he misses a question. What is the probability that he will miss on the ath question?"

The solution [14] is easily shown to be

$$\text{pr}_k(a) = \frac{(N-k)!(N-a)!k}{(N-k-a+1)!N!}.$$ (27)

In particular, $\text{pr}_1(a) = 1/N$, agreeing with Theorem 1; and $\text{pr}_2(a) = 2(N-a)/N(N-1)$, which decreases linearly with a. Properties of the model Q which go deeper than Eq. (27) do not appear to be directly related to the shift register cycle problem.

4. ALTERNATIVE APPROACHES TO THE DISTRIBUTION OF THE LONGEST CYCLE

4.1. *The differential difference equation method*

In the previous discussion, the expected relative length of the longest cycle, λ_N, was defined for random permutations on N objects. Properties of this sequence were obtained, and upper and lower bounds were computed for $\lim_{N \to \infty} \lambda_N$. In the present section, the limiting distri-bution of the relative length of the longest cycle is represented as the solution of a simple differential difference equation. This equation permits a machine attack on the problem, and the first four moments were computed with an estimated error of less than 10^{-8}:

$$m_1 = 0.62432965$$
$$m_2 = 0.42669557$$
$$m_3 = 0.31363052$$
$$m_4 = 0.24387645$$

Let $X_1^{(N)}, X_2^{(N)}, X_3^{(N)}, \ldots$ be the sequence of cycle lengths obtained from the N-object statistical model of Section 3, and let

$$Z_N = \max [X_1^{(N)}, X_2^{(N)}, X_3^{(N)}, \ldots].$$

Now

$$Z_N = \max [X_1^{(N)}, \max (X_2^{(N)}, X_3^{(N)}, \ldots)],$$

and therefore Z_N has the same distribution as

$$\max [X_1^{(N)}, Z_M]$$

where $M = N - X_1^{(N)}$, since the conditional distribution of $X_2^{(N)}, X_3^{(N)}, \ldots$, given $X_1^{(N)}$, is the same as the distribution of $X_1^{(M)}, X_2^{(M)}, X_3^{(M)}, \ldots$.

Letting

$$F_N(\alpha) = \mathrm{pr} [Z_N \leq N\alpha],$$

we have

$$F_N(\alpha) = \mathrm{pr} [(X_1^{(N)} \leq N\alpha) \cap (Z_M \leq N\alpha)]$$
$$= [\mathrm{pr} (X_1^{(N)} \leq N\alpha)][\mathrm{pr} (Z_M \leq N\alpha)]$$
$$= \sum_{x=1}^{[N\alpha]} \left[\frac{1}{N}\right] \left[F_{N-x}\left(\frac{N\alpha}{N-X}\right)\right], \tag{28}$$

where $[N\alpha]$ denotes the greatest integer $\leq N\alpha$.

Next, set $x = X/N$. Bypassing the question of convergence, but taking the limit as $N \to \infty$, we find[1]

$$F(\alpha) = \int_0^\alpha F\left(\frac{\alpha}{1-x}\right) dx = \alpha \int_\alpha^{\alpha/(1-\alpha)} \frac{F(z)}{z^2} dz. \tag{29}$$

The derivative is

---

[1] $F_N(\alpha)$ can be shown to converge in the range $1/2 \leq \alpha \leq 1$, and then by in-duction $F_N(\alpha)$ can be shown to converge in the range $1/(m+1) \leq \alpha \leq 1/m$. The convergence of F_N implies that Eq. (29) follows from Eq. (28).

$$F'(\alpha) = \int_\alpha^{\alpha/(1-\alpha)} \frac{F(z)}{z^2}\,dz - \frac{F(\alpha)}{\alpha} + \frac{F\left(\dfrac{\alpha}{1-\alpha}\right)}{\alpha}$$

$$= \frac{F(\alpha)}{\alpha} - \frac{(F\alpha)}{\alpha} + \frac{F\left(\dfrac{\alpha}{1-\alpha}\right)}{\alpha} \tag{30}$$

$$= \frac{F\left(\dfrac{\alpha}{1-\alpha}\right)}{\alpha}.$$

Define

$$R(Y) = \frac{1}{Y}F'\left(\frac{1}{Y}\right) = F\left(\frac{1}{Y-1}\right) \qquad 1 \le Y < \infty.$$

Then

$$R'(Y) = -\frac{F'\left(\dfrac{1}{Y-1}\right)}{(Y-1)^2} = -\frac{R(Y-1)}{Y-1} \quad \text{for} \quad Y \ge 2 \tag{31}$$

and

$$R(Y) \equiv 1 \quad \text{for} \quad 1 \le Y \le 2 \tag{32}$$

since, in this region $R(Y) = F(1/(Y-1)) = \lim_{N \to \infty} F_N(1/(Y-1)) = \lim_{N \to \infty} \text{pr}[Z_N \le N/(Y-1)] = 1$, in view of $Z_N \le N \le N/(Y-1)$. Equation (31) is the differential difference equation for the distribution of the longest cycle. Its solution can be obtained, in principle, by successive integrations for successive unit intervals along the Y-axis. Thus:

$$R(Y) = 1 \quad \text{for } 1 \le Y \le 2$$
$$R(Y) = 1 - \log(Y-1) \quad \text{for } 2 \le Y \le 3$$
$$R(Y) = -\int^Y \frac{1 - \log(t-1)}{t-1}\,dt \quad \text{for } 3 \le Y \le 4,$$
$$\text{etc.}$$

The moments of the maximum cycle length are then given by

$$\int_0^1 \alpha^n\,dF(\alpha) = \int_1^\infty \frac{1}{Y^{n+1}} R(Y)\,dY, \tag{33}$$

which was used in the computer determination of the first four moments, as listed at the start of this section.

4.2. *Expression by definite integrals*

In the November, 1964, issue of the *Bulletin of the American Mathematical Society*, the determination of the constant "λ" was proposed as a Research Problem. In response to this problem, several mathematicians succeeded in finding definite integral representations for λ. Three such expressions are:

1. $\qquad \lambda = \int_0^\infty e^{-t - \mathrm{Ei}(t)} dt$, where $\mathrm{Ei}(t) = \int_t^\infty \frac{e^{-u}}{u} du$,

discovered by Lawrence A. Shepp of the Bell Telephone Laboratories.

2. $\qquad \lambda = \int_0^\infty (1 - e^{-\int_0^\infty (e^{-t}/t) dt}) ds$,

contributed by Walter Weissblum, Michael Lieber, and Jack Warga, all of AVCO Corporation.

3. $\qquad \lambda = \int_0^1 e^{\mathrm{li}(x)} dx$, where $\mathrm{li}(x) = \int_0^x \frac{dt}{\log t}$,

from Nathan Fine of Pennsylvania State College.

The equivalence of these three expressions is an exercise in the integral calculus. The question of whether λ is rational, irrational algebraic, or transcendental still remains unanswered.

5.   SHIFT REGISTER CYCLES OF ALL LENGTHS

From an n-stage shift register with nonlinear feedback, it is possible to obtain various cycles of binary digits, with periods ranging between 1 and 2^n. Recently, the question was posed whether *all* periods between 1 and 2^n inclusive can be so obtained, using shift registers of degree n. In the present section this question will be answered in the affirmative. Moreover, given a maximum-length *linear* shift register (period $2^n - 1$), every cycle period between 1 and 2^n can be obtained from it by one of 2^{n-1} simple, standard modifications. (The number of necessary modifications is only half the number of cycle lengths obtained, because a given shift register with feedback yields more than a single cycle length, in general.)

5.1.   *The existence proof*

For every n, there are linear shift-register sequences of period $2^n - 1$. (Precisely, there are $\phi(2^n - 1)/n$ such sequences for each n.) Let $\{a_k\}$ be such a sequence. To find a cycle of length L, for $1 \leq L \leq 2^n$, consider the modulo-2 sum $\{a_k\} \oplus \{a_{k+L}\}$. By the "delay-and-add" property of maximum-length linear sequences, for some integer M, $\{a_k\} \oplus \{a_{k+L}\} = \{a_{k+M}\}$, where $\{a_{k+M}\}$ is merely a new phase shift of the original maximum-length sequence. Thus, every set of n consecutive bits, except n 0's, occurs in $\{a_{k+M}\}$. In particular, for some k_0, we have

$a_{k_0+M+1} = a_{k_0+M+2} = \cdots = a_{k_0+M+n-1} = 0$, $a_{k_0+M+n} = 1$; that is, we find $00\ldots01$ somewhere in the sequence. This in turn implies that

$$a_{k_0+1} = a_{k_0+L+1}$$
$$a_{k_0+2} = a_{k_0+L+2}$$
$$\vdots$$
$$a_{k+n-1} = a_{k_0+L+n-1}$$
$$a_{k_0+n} = a_{k_0+L+n} \oplus 1$$

Thus we have found two n-bit subsequences of $\{a_k\}$, spaced L positions apart, which agree in their first $n-1$ positions, but disagree in the nth position. This easily allows a modification of the feedback logic to produce a cycle of length L. Specifically, the logical function (a large *and*-gate) is constructed which has a 1 output if and only if it detects the values of $a_{k_0+1}, a_{k_0+2}, \ldots, a_{k_0+n-1}$ in the front $n-1$ positions of the shift register. The output of this *and*-gate is added modulo 2 to the original feedback function, which has the effect of splitting the original cycle, of period $2^n - 1$, into two cycles of periods L and $2^n - L - 1$, respectively. This argument is valid for all periods from 1 to $2^n - 2$, inclusive. The original sequence itself has period $2^n - 1$, and there are many nonlinear sequences ($2^{2^{n-1}-n}$ in all) of length 2^n, including one obtained from the linear sequence using an *and*-gate that looks for $n-1$ *zeros*.

5.2. *Constructive examples*

For $n = 4$, the linear feedback logic $a_k = a_{k-3} \oplus a_{k-4}$ yields a cycle of length $2^n - 1 = 15$. There are eight terms of the type $a_{k-1}^r a_{k-2}^s a_{k-3}^t$, where r, s, and t are binary exponents which indicate complementation (when they are 1) or no complementation (when they are 0). The modified feedback logic

$$a_k = a_{k-1}^r a_{k-2}^s a_{k-3}^t \oplus a_{k-3} \oplus a_{k-4},$$

in accordance with the existence proof, will yield any desired cycle length from 1 to 16 by suitable specification of r, s, and t, except length 15, which, however, is already available from the original linear sequence. The results obtained for all combinations of r, s, and t are shown in Table VII-8. (In all cases but one, the cycle $00\ldots0$ occurs separately, so that the appearance of 1 as a cycle length is not unique, as are the appearances of all other lengths.)

For large n, it would be reasonable to search through $\{a_k\} \oplus \{a_{k+L}\}$ for given L, using a shift register, a delay line, and a word detector, to determine the instant at which $(0, 0, \ldots, 0, 1)$ occurs as the sum. A word detector, built to detect the first $n-1$ digits which appeared

Table VII-8.   Cycle lengths obtained from the shift register
$$a_k = a_{k-1}^r a_{k-2}^s a_{k-3}^t \oplus a_{k-3} \oplus a_{k-4}.$$

r	s	t	Cycle Lengths		
0	0	0	1	1	14
0	0	1	1	5	10
0	1	0	1	2	13
0	1	1	1	3	12
1	0	0	1	7	8
1	0	1	1	6	9
1	1	0	1	4	11
1	1	1	16		

in the $\{a_k\}$ register at that instant, can then be appended, modulo 2, to the feedback logic of the shift register, so that the modified sequence has period L.

5.3. *A false conjecture*

The following plausible extension of the result that every cycle length occurs can be proven false:

Conjecture B. The number of cycles of length L obtainable from *all* n-stage permutation shift registers is a positive integral multiple of $B_n = 2^{2^{n-1}}/2^n$ for $1 \leq L \leq 2^n$.

The number B_n was shown by de Bruijn to be the exact number of permutation shift registers from which cycles of maximum length, $L = 2^n$, are obtainable. It is easily shown that the number of cycles of length $L = 2^n - 1$ obtained from all $2^{2^{n-1}}$ permutation shift registers is exactly $2B_n$; and at the other end, Conjecture B holds for $L = 1$ (with $2^{2^{n-1}}$ one-cycles in all), and for $L = 2$ (with $\frac{1}{2}2^{2^{n-1}}$ two-cycles). However, for $L = 2^n - 2$, the number of cycles of length L is strictly between B_n and $2B_n$ for all $n > 3$. In particular, cycles of length $2^n - 2$ arise from cycles of length 2^n either by removal of the two one-cycles (all 0's and all 1's), or by removal of the unique two-cycle (alternating 0's and 1's). There are B_n instances of removal of both one-cycles, but the number of instances of removal of the two-cycle is strictly between 0 and B_n for $n > 3$. This disproves Conjecture B.

A closely related question, also unanswered, is: What is the total number of cycles obtained from all $2^{2^{n-1}}$ permutation shift registers?

6.   THE EXPERIMENTAL APPROACH

The purpose of the computer program was the experimental de-

termination of the length and number of all cycles associated with a
given recursion formula for a given register length. The circuitry was
built up with modules of digital logic elements. These modules could
be interconnected to form shift registers, counters, coincidence gates,
and other basic digital building blocks. All signals consisted of pulses,
at a megacycle rate, synchronized to a master clock.

The recursion formula was mechanized by the usual technique of
an n-stage shift register with a nonlinear feedback logic. In order to
investigate every cycle all of the 2^n possible initial words were inserted,
in turn, into the shift register. Thus started upon a cycle, the shift
register continued to compute new words until the word initially inserted returned. The cycle length equals the amount of time required
(in clock intervals) for the complete recirculation of the inserted word.
This time was measured by an external counter and was recorded on
the paper tape of a printout device. The number of times that each

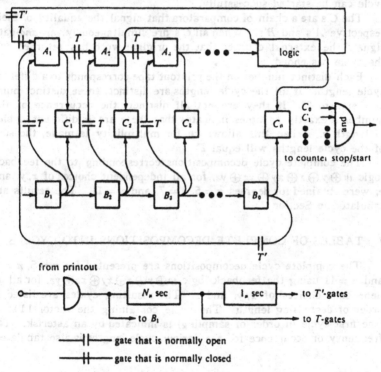

Fig. VII-4. Block diagram of the computer used to generate all
cycles for a given shift register.

cycle was represented on the tape was equal to the number of words contained in that cycle.

A block diagram of the device is shown in Figure VII-4. The A's comprise a shift register, which, when combined with the logic, gener-ates the recursive sequence. The B's are dynamic flip-flops that count once for each input pulse. The total count on the flip-flops is the word that is to be inserted.

When a pulse arrives from the printout device, signaling readiness to commence operation, the total on the counters is increased by one. After a delay (to allow all counters to reach equilibrium) the new word is inserted into the shift register. Insertion is accomplished in two steps. First, all of the T-gates are energized; this transfers B_1 through B_n into the shift register. One clock interval later the T'-gates are energized, transferring B_0 into A_1. Two steps are needed as the logic has 1 μsec of storage, and it is necessary to clear this before a new cycle can be started successfully.

The C's are a chain of comparators that signal the equality of their respective A's and B's. When all C's fire simultaneously, an *and*-gate signals the external counter that the original word has returned and the cycle has ended.

Each distinct number on the printout tape corresponds to a different cycle length. If all the cycle lengths are distinct, these distinct num-bers sum to 2^n. If they are not all distinct, the occurrence of the number q exactly pq times indicates that there are p different cycles of length q. When this allowance for multiplicity is made, the sum of the cycle lengths will equal 2^n.

The complete cycle decompositions corresponding to the feedback logic $w \oplus y \oplus z \oplus xy \oplus xz \oplus yz$, for all independent choices of x, y, and z, were obtained for degrees $n = 5$, $n = 7$, and $n = 11$. The results are tabulated in Section 7.

7. TABLES OF COMPLETE DECOMPOSITIONS INTO CYCLES

The complete cycle decompositions are presented for $n = 5$, $n = 7$, and $n = 11$ using the feedback logic $w \oplus y \oplus z \oplus xy \oplus xz \oplus yz$, for all in-dependent choices of x, y, and z. The resulting cycles are listed in order of decreasing length. The cycle containing the vector $111 \ldots 1$ (the first cycle in order of sampling) is indicated by an asterisk. The frequency of occurrence for the various cycle lengths is also tabulated.

Table VII-9. Lengths of cycles, for $n = 5$.

x	y	z	Cycle Lengths			
1	2	3	24*	4	3	1
1	2	4	13*	10	8	1
1	3	4	31*	1		
2	1	3	31*	1		
2	1	4	24*	4	3	1
2	3	4	31*	1		

* Cycle containing the vector 111...1.

Table VII-10. Frequency of occurrence of cycle lengths, for $n = 5$.

Cycle Length	Frequency	Cycle Length	Frequency	Cycle Length	Frequency
1	6	12	0	22	0
2	0	13	1	23	0
3	2	14	0	24	2
4	2	15	0	25	0
5	0	16	0	26	0
6	0	17	0	27	0
7	0	18	0	28	0
8	1	19	0	29	0
9	0	20	0	30	0
10	1	21	0	31	3
11	0				

Total 18

Table VII-11. Lengths of cycles, for $n = 7$.

x	y	z	Cycle Lengths							
1	2	3	127*	1						
1	2	4	90*	34	3	1				
1	2	5	99*	12	8	5	3	1		
1	2	6	86	22	19*	1				
1	3	4	81*	26	11	5	4	1		
1	3	5	106*	11	10	1				
1	3	6	39	30	27*	20	11	1		
1	4	5	115*	9	3	1				
1	4	6	79*	26	22	1				
1	5	6	62	33	32*	1				
2	1	3	58	39	17*	10	3	1		
2	1	4	127*	1						
2	1	5	104*	15	8	1				
2	1	6	52*	32	13	13	8	6	3	1

Table VII-11 (Cont'd).

x	y	z	Cycle Lengths							
2	3	4	120*	4	3	1				
2	3	5	51	39	37*	1				
2	3	6	101*	14	12	1				
2	4	5	127*	1						
2	4	6	56	37*	16	15	3	1		
2	5	6	127*	1						
3	1	2	113*	10	4	1				
3	1	4	127*	1						
3	1	5	68*	37	22	1				
3	1	6	71*	33	13	6	4	1		
3	2	4	101	18*	8	1				
3	2	5	62	53*	5	4	3	1		
3	2	6	79*	45	3	1				
3	4	5	77*	42	8	1				
3	4	6	113	8	6	1				
3	5	6	54	52*	14	4	3	1		

* Cycle containing the vector 111...1.

Table VII-12. Frequency of occurrence of cycle lengths, for $n = 7$.

Cycle Length	Frequency	Cycle Length	Frequency	Cycle Length	Frequency
1	30	19	1	58	1
2	0	20	1	62	2
3	10	22	3	68	1
4	6	26	2	71	1
5	3	27	1	77	1
6	3	30	1	79	2
7	0	32	2	81	1
8	6	33	2	86	1
9	1	34	1	90	1
10	3	37	3	99	1
11	3	39	3	101	2
12	2	42	1	104	1
13	3	45	1	106	1
14	2	51	1	113	2
15	2	52	2	115	1
16	1	53	1	120	1
17	1	54	1	127	5
18	1	56	1		

Total   130

Table VII-13. Lengths of cycles, for $n = 11$.

x	y	z	Cycle Lengths
1	2	3	818* 816 342 39 15 14 3 1
		4	1934* 70 43 1 1
		5	1097 855* 95 1
		6	1650* 260 126 8 3 1
		7	934 886* 227 1
		8	728* 547 303 273 133 36 27 1
		9	872 617 177* 139 89 36 24 20 20 13 10 9 9 9 3 1
		10	1683 241 70 31* 22 1
1	3	4	1460 220 162 93* 80 28 4 1
		5	1057 731* 123 54 52 27 3 1
		6	1609* 403 35 1
		7	1513 484* 50 1
		8	850 285 225* 225 123 104 31 19 19 19 19 19 19 14 14 14 8 4 3 1
		9	414 299 245 244* 243 143 141 106 93 43 37 25 14 1
		10	1380* 361 162 65 51 14 14 1
1	4	5	1276 577* 64 52 50 19 9 1
		6	1906* 119 22 1
		7	669 489 223 163 150 114* 114 50 21 21 12 7 7 4 3 1
		8	1156* 303 180 110 66 61 44 22 22 22 22 22 17 1
		9	953 810* 154 83 47 1
		10	1338* 446 156 52 39 13 3 1

* Cycle containing the vector 111...1.

Table VII-13 (Cont'd).

x	y	z	Cycle Lengths																				
1	5	6	1639*	211	96	17	17	17	17	17	16	8	6	3	1								
		7	929*	652	198	172	27	22	17	17	16	14	1									
		8	1264	395*	170	122	34	26	17	12	7	1										
		9	870*	852	153	112	47	10	3	1												
		10	1110*	430	399	76	32	1														
1	6	7	1349*	390	125	103	52	19	9	1												
		8	1021	393	364*	155	58	30	16	7	3	1										
		9	1168	734*	145	1																
		10	745	611*	504	100	42	37	8	1												
1	7	8	626	585	288*	235	161	52	49	47	4	1										
		9	629*	430	370	327	196	74	21	1	1											
		10	912*	912	75	60	60	25	3	1	1											
1	8	9	736*	588	320	174	137	50	29	10	3	1										
		10	1051*	626	167	65	48	42	20	15	13	1										
1	9	10	1141	288	270	191*	86	38	33	1												
2	1	3	1099	850*	52	39	7	1														
		4	1765*	205	61	13	3	1														
		5	829	569	407	165*	45	16	8	5	3	1										
		6	1829*	177	41	1																
		7	1102	507	312*	93	16	14	3	1												
		8	2009*	35	3	1																
		9	819*	736	375	43	26	25	23	1												
		10	592*	561	389	159	139	26	21	21	21	21	21	21	21	21	10	10	10	7	5	3	1

2	3	4	1182*	371	229	90	84	42	31	14	14	8	7	4	1			
		5	868*	661	253	211	32	16	6	1	1							
		6	1756*	129	53	40	33	27	9	1	1							
		7	643	568*	546	140	107	22	21	1	1							
		8	1664*	118	89	53	19	19	19	19								
		9	1346	683*	18	1												
		10	1701	324*	22	1												
2	4	5	1201	233*	225	207	93	85	3	1	1							
		6	764	732*	209	140	77	63	62	1	1							
		7	1862*	54	41	18	18	18	15	7	7	4	3	1				
		8	1687	94	85	66*	37	27	26	22	3	1						
		9	1247*	718	82	1	1											
		10	2030*	14	3	1	1											
2	5	6	635	532	464*	159	53	52	40	27	17	17	16	16	8	6	5	1
		7	963	831	235*	15	3	1										
		8	1150	324	226	191	117*	29	10	1	1							
		9	1986*	40	8	7	6	1										
		10	1965*	79	3	1	1											
2	6	7	1972*	59	16	1												
		8	920*	906	117	67	37	1										
		9	1769*	172	57	42	7	1										
		10	1542	358*	81	23	21	17	5	1								

* Cycle containing the vector 111...1.

Table VII-13 (Cont'd).

x	y	z				Cycle Lengths									
2	7	8	1056	663*	246	42	33	4	3	1					
		9	1193	745*	109	1									
		10	1908*	136	3	1									
2	8	9	1250	392*	95	81	64	59	50	30	26	1			
		10	659	559*	309	211	147	113	39	7	3	1			
2	9	10	1308	523	216*	1									
3	1	2	1047	562	273	161*	4	1							
		4	906	385	263	182*	104	70	51	41	26	3	1		
		5	874	663	436*	69	5	1							
		6	1076	845*	78	22	19	4	3	1					
		7	1185*	284	223	179	135	38	3	1					
		8	1684*	174	90	52	47	1							
		9	1343*	576	97	28	3	1							
		10	1049*	665	238	21	21	21	10	10	5	4	3	1	
3	2	4	675*	642	360	207	94	64	5	1					
		5	1692	198*	137	16	4	1							
		6	919*	321	232	171	158	112	91	37	6	1			
		7	1070*	497	214	165	50	30	21	1					
		8	931	511	181	128	118	79*	51	32	16	1			
		9	921*	461	251	163	87	58	39	20	20	9	9	5	4
		10	1393	412*	124	98	20	1							

3	4	5	1070	558*	194	124	51	32	18	18	8	7	5	3	1
		6	683*	662	333	280	53	33	33	3	1	7	3		
		7	672	557*	443	177	114	18	18	18	1				
		8	620	426*	332	291	185	158	35	1					
		9	1184*	860	3	1	3	1							
		10	1605*	296	130	13	3								
3	5	6	1051*	506	301	79	48	17	16	14	6	5	4	1	
		7	1713*	317	17	1									
		8	2047*	1											
		9	1096*	418	405	53	39	26	10	1					
		10	1487*	462	87	7	4	1							
3	6	7	1878*	97	60	9	3	1	1	1					
		8	1579*	366	31	27	22	16	6						
		9	1856*	105	50	32	4	1							
		10	905	759*	217	68	59	24	7	5	3	1			
3	7	8	986	731*	187	123	20	1	22	17	16	5	3	1	
		9	674*	450	400	335	63	62	22	17	16				
		10	1183	494	308*	59	3	1							
3	8	9	1385*	336	326	1									
		10	1111*	437	185	164	95	39	16	1					
3	9	10	1077	396*	258	153	91	36	29	4	3	1			

* Cycle containing the vector 111...1.

Table VII-13 (Cont'd).

Cycle Lengths

x	y	#																			
4	1	2	1108*	646	195	37	35	22	4	1											
		3	834	633	234	126	64	54	48	47*	7	1									
		5	1064*	667	197	53	46	15	5	1											
		6	990*	342	287	163	135	64	53	9	4	1									
		7	882	882	186*	62	24	8	3	1											
		8	1237*	540	132	106	30	1													
		9	1014*	696	245	69	23	1													
		10	885	360*	360	295	30	30	21	15	12	10	10	7	5	4	3	1			
4	2	3	1692*	110	87	69	50	36	3	1											
		5	1183	755*	46	30	22	7	4	1											
		6	1361	356*	177	150	3	1													
		7	1622*	314	73	30	8	1													
		8	526	520	519	302	152*	18	10	1											
		9	1130*	468	119	84	49	30	20	20	20	20	13	10	10	9	9	9	4	3	1
		10	1028	862*	90	47	20	1													
4	3	5	1659*	286	78	21	3	1													
		6	1888*	55	50	22	15	10	7	1											
		7	1095*	646	157	53	47	33	16	1											
		8	1154*	330	239	152	59	30	19	19	19	8	8	7	3	1					
		9	1522*	326	199	1															
		10	959	735	205*	110	38	1													

a	b	c																
4	5	6	842*	739	223	173	52	6	5	4	3	1					
4	5	7	1693*	346	8	1	37	19	7	3	1						
4	5	8	1757*	235	55	1											
4	5	9	1676*	103	76	65	61	22	22	22	22	4	1				
4	5	10	1668*	150	47	44	24	22	22	22	22	3	1				
4	6	7	1664*	369	14	1											
4	6	8	1053	569	391*	31	3										
4	6	9	1976*	67	4	1											
4	6	10	1153	458*	102	86	77	68	45	37	9	7	5	1			
4	7	8	1358*	577	79	25	8	1									
4	7	9	954	673	165	142	69*	36	8	1							
4	7	10	636*	489	285	212	163	95	66	66	24	8	3	1			
4	8	9	969	941*	64	37	20	13	3	1							
4	8	10	1518	369*	118	35	7	1									
4	9	10	1753*	165	101	24	4	1									
5	1	2	1717*	270	49	8	3	1	1								
5	1	3	1348	456*	142	82	19	1	1								
5	1	4	1297	389*	237	88	20	13	3	1							
5	1	6	1206	486*	355	1											
5	1	7	1268	776*	3	1											
5	1	8	1204	840*	3	1											
5	1	9	1252*	451	201	66	32	28	17	1							
5	1	10	1197	471*	117	72	53	21	21	21	21	12	10	10	10	8	3	1

* Cycle containing the vector 111...1.

Table VII-13 (Cont'd).

x	y	z	Cycle Lengths															
5	2	3	1075*	370	366	214	22	1										
		4	777*	402	322	316	211	16	3	1								
		6	2047*	1														
		7	1264*	741	21	18	3	1										
		8	1971*	66	10	1												
		9	900*	450	235	146	74	70	20	20	20	20	13	13	9	9	8	1
		10	841*	566	546	58	17	16	3	1								
5	3	4	1993*	50	4	1	1											
		6	1599*	334	63	32	19	1										
		7	1733*	256	58	1	1											
		8	1185	771*	19	19	19	14	8	8	4	1						
		9	1107*	498	333	94	15	1										
		10	1174*	294	226	161	60	54	31	30	17	1						
5	4	6	868*	571	241	220	147	1	1									
		7	1409*	324	105	94	50	18	18	15	7	4	3	1				
		8	1131	550	137	98	69	59*	3	1								
		9	1218*	551	142	83	53	1	3	1								
		10	711*	579	263	262	216	13	3	1								
5	6	7	1418*	418	75	70	66	1										
		8	1459*	384	136	37	31	1										
		9	1530*	332	118	41	26	1										
		10	1232	376	195	174*	38	22	10	1								

5	7	8	1798*	155	87	4	3	1			
		9	1956*	66	25	1					
		10	1236*	599	87	81	25	16	3	1	
5	8	9	769*	381	374	373	76	39	35	1	
		10	2023*	21	3	1					
5	9	10	1323	531*	88	38	37	22	8	1	

* Cycle containing the vector 111...1.

Table VII-14. Frequency of occurrence of cycle lengths, for $n = 11$.

Cycle Length	Frequency	Cycle Length	Frequency	Cycle Length	Frequency	Cycle Length	Frequency
1	180	39	8	77	2	118	4
2	0	40	3	78	2	119	2
3	66	41	4	79	4	122	1
4	30	42	5	80	1	123	3
5	16	43	3	81	3	124	2
6	8	44	2	82	2	125	1
7	25	45	2	83	2	126	2
8	23	46	2	84	2	128	1
9	16	47	8	85	2	129	1
10	22	48	3	86	2	130	1
11	0	49	3	87	5	132	1
12	4	50	11	88	2	133	1
13	11	51	4	89	2	135	2
14	17	52	9	90	3	136	2
15	9	53	10	91	2	137	3
16	20	54	4	92	0	139	2
17	16	55	2	93	4	140	2
18	12	56	0	94	4	141	1
19	23	57	1	95	4	142	3
20	21	58	4	96	1	143	1
21	23	59	6	97	2	145	1
22	25	60	4	98	2	146	1
23	3	61	3	99	0	147	2
24	6	62	3	100	1	150	3
25	6	63	3	101	1	152	2
26	8	64	6	102	1	153	2
27	7	65	3	103	2	154	1
28	3	66	8	104	2	155	2
29	3	67	2	105	2	156	1
30	11	68	2	106	1	157	1
31	7	69	5	107	1	158	2
32	7	70	5	108	1	159	2
33	5	71	0	109	1	161	3
34	1	72	1	110	3	162	2
35	6	73	1	112	2	163	4
36	5	74	2	113	1	164	1
37	11	75	2	114	3	165	4
38	5	76	3	117	3	167	1

Table VII-14 (Cont'd).

Cycle Length	Frequency	Cycle Length	Frequency	Cycle Length	Frequency	Cycle Length	Frequency
170	1	235	4	320	1	399	1
171	1	237	1	321	1	400	1
172	2	238	1	322	1	402	1
173	1	239	1	324	3	403	1
174	3	241	2	326	2	405	1
177	4	243	1	327	1	407	1
179	1	244	1	330	1	412	1
180	1	245	2	332	2	414	1
181	1	246	1	333	2	418	2
182	1	251	1	334	1	426	1
185	2	253	1	335	1	430	2
186	1	256	1	336	1	436	1
187	1	258	1	342	2	437	1
191	2	260	1	346	1	443	1
194	1	262	1	355	1	446	1
195	2	263	2	356	1	450	2
196	1	270	2	358	1	451	1
197	1	273	2	360	3	456	1
198	2	280	1	361	1	458	1
199	1	284	1	364	1	461	1
201	1	285	2	366	2	462	1
205	2	286	1	369	2	464	1
207	2	287	1	370	2	468	1
209	1	288	2	371	1	471	1
211	4	291	1	373	1	484	1
212	1	294	1	374	1	486	1
214	2	295	1	375	1	489	2
216	2	296	1	376	1	494	1
217	1	299	1	381	1	497	1
220	2	301	1	384	1	498	1
223	3	302	1	385	1	504	1
225	3	303	2	389	2	506	1
226	2	308	1	390	1	507	1
227	1	309	1	391	1	511	1
229	1	312	1	392	1	519	1
232	1	314	1	393	1	520	1
233	1	316	1	395	1	523	1
234	1	317	1	396	1	526	1

Table VII-14 (Cont'd).

Cycle Length	Fre-quency	Cycle Length	Fre-quency	Cycle Length	Fre-quency	Cycle Length	Fre-quency
531	1	663	2	845	1	1056	1
532	1	665	1	850	2	1057	1
540	1	667	1	852	1	1064	1
546	2	669	1	855	1	1070	2
547	1	672	1	860	1	1075	1
550	1	673	1	862	1	1076	1
551	1	674	1	868	2	1077	1
557	1	675	1	870	1	1095	1
558	1	683	2	872	1	1096	1
559	1	696	1	874	1	1097	1
561	1	711	1	882	2	1099	1
562	1	718	1	885	1	1102	1
566	1	728	1	886	1	1107	1
568	1	731	2	900	1	1108	1
569	2	732	1	905	1	1110	1
571	1	734	1	906	2	1111	1
576	1	735	1	912	2	1130	1
577	2	736	2	919	1	1131	1
579	1	739	1	920	1	1141	1
585	1	741	1	921	1	1150	1
588	1	745	2	929	1	1153	1
592	1	755	1	931	1	1154	1
599	1	759	1	934	1	1156	1
611	1	764	1	941	1	1168	1
617	1	769	1	953	1	1174	1
620	1	771	1	954	1	1182	1
626	2	776	1	959	1	1183	2
629	1	777	1	963	1	1184	1
633	1	810	1	969	1	1185	2
635	1	816	1	986	1	1193	1
636	1	818	1	990	1	1197	1
642	1	819	1	1014	1	1201	1
643	1	829	1	1021	1	1204	1
646	2	831	1	1028	1	1206	1
652	1	834	1	1047	1	1218	1
659	1	840	1	1049	1	1232	1
661	1	841	1	1051	2	1236	1
662	1	842	1	1053	1	1237	1

Table VII-14 (Cont'd).

Cycle Length	Frequency	Cycle Length	Frequency	Cycle Length	Frequency	Cycle Length	Frequency
1247	1	1409	1	1668	1	1862	1
1250	1	1418	1	1676	1	1878	1
1252	1	1459	1	1683	1	1888	1
1264	2	1460	1	1684	1	1906	1
1268	1	1487	1	1687	1	1908	1
1276	1	1513	1	1692	2	1934	1
1297	1	1518	1	1693	1	1956	1
1308	1	1522	1	1701	1	1965	1
1323	1	1530	1	1713	1	1971	1
1338	1	1542	1	1717	1	1972	1
1343	1	1579	1	1733	1	1976	1
1346	1	1599	1	1753	1	1986	1
1348	1	1605	1	1756	1	1993	1
1349	1	1609	1	1757	1	2009	1
1358	1	1622	1	1765	1	2023	1
1361	1	1639	1	1769	1	2030	1
1380	1	1650	1	1798	1	2047	2
1385	1	1659	1	1829	1	Total	1408
1393	1	1664	2	1856	1		

Chapter VIII

ON THE CLASSIFICATION OF BOOLEAN FUNCTIONS

1. INTRODUCTION

The Boolean functions of k variables, $f(x_1, x_2, \ldots, x_k)$, fall into equivalence classes (or *families*) when two functions differing only by permutation or complementation of their variables are considered *equivalent*. The number of such families is easily computed, as illustrated by Slepian [36]. The next step is to discover the *invariants* of the logic families, and determine to what extent they characterize the individual families. Given the class decomposition, one also wishes to select a "representative assembly" with one delegate from each family. That is, *canonical forms* for the logics are sought, with every family having its characteristic canonical form. This mathematical program will be carried out in Sections 2 through 6.

Given certain of the invariants, it is possible to say something about the size of the corresponding family. Applications of this principle are explored in Section 7.

The practical significance of the symmetry classes is that a circuit which mechanizes a given function f will also mechanize any other function of the same class, provided that complements of all the inputs are available, simply by permuting and complementing the inputs. In particular, the extensive investigations on the minimization of logical circuitry can be confined, without loss of generality, to one representa-

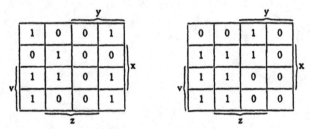

Fig. VIII-1. Two Boolean functions belonging to the same symmetry class.

tive function in each symmetry class.

Example. The two Karnaugh charts of Figure VIII-1 represent Boolean functions belonging to the same symmetry class. In particular, if the function at the left is designated $f(v, x, y, z)$, then the function on the right is $f(x, v', z', y)$.

2. THE ZERO AND FIRST ORDER INVARIANTS

Definition. Two Boolean functions $f_1(x_1, x_2, \ldots, x_k)$ and $f_2(x_1, x_2, \ldots, x_k)$ will be called *equivalent* if there is a way of permuting and complementing the variables $x_1, x_2, \ldots, x_k$ to produce variables $y_1, y_2, \ldots, y_k$ such that

$$f_1(y_1, y_2, \ldots, y_k) \oplus f_2(x_1, x_2, \ldots, x_k) = b \,,$$

b a binary constant, for all 2^k assignments of values to the position vector $(x_1, x_2, \ldots, x_k)$.

Note: The group G of allowable symmetries contains $k!$ permutations of the variables, 2^k complementations of the variables, and a twofold choice of the constant b. Thus G consists of $2^{k+1} \cdot k!$ operators. In particular, the size of any family of equivalent Boolean functions is some *factor* of $2^{k+1} \cdot k!$ Since there are 2^{2^k} Boolean functions of k variables, the number of families is approximately $(2^{2^k})/(2^{k+1} \cdot k!)$.

Theorem 1. The quantity $T_0 = \max(c_0, 2^k - c_0)$, where $c_0 = \sum_{x_1, x_2, \ldots, x_k} f(x_1, x_2, \ldots, x_k)$, is an *invariant* for the family containing $f(x_1, x_2, \ldots, x_k)$. (The capital sigma denotes ordinary summation.)

Proof. The quantity c_0 is the sum of all the entries in the truth table for $f(x_1, x_2, \ldots, x_k)$. Permutation and complementation of the variables *rearranges* the entries in the truth table, but leaves their sum c_0 unaltered. Complementing $f(x_1, x_2, \ldots, x_k)$ replaces c_0 by $2^k - c_0$, which means that half the operations of G leave c_0 fixed, and the other half replace c_0 by $2^k - c_0$. Thus $T_0 = \max(c_0, 2^k - c_0)$ is invariant under *all* operations of G.

> *Corollary.* Relative to the subgroup H of G which allows permutation and complementation of the variables but forbids complementation of the function (so that $[G/H] = 2$), the quantity c_0 is itself an invariant of the equivalence classes.

Definition. T_0 is the *zero order invariant* of the logic family for the group G. (Likewise c_0 is the *zero order invariant* for the group H.)

Theorem 2.   For every i, $1 \leqq i \leqq k$, let $R_1^i = \max(c_1^i, 2^k - c_1^i)$, where $c_1^i = \sum\limits_{x_1, x_2, \ldots, x_k} (f(x_1, x_2, \ldots, x_k) \oplus x_i)$. Then the set of numbers $R_1^1, R_1^2, \ldots, R_1^k$, when rearranged in descending order, forms a collection of k invariants $T_1^1, T_1^2, \ldots, T_1^k$ for the family to which f belongs. (The symbol $\oplus$ denotes modulo 2 addition.)

Proof.   The number R_1^i is not changed when x_i is replaced by its complement, since $\max(c_1^i, 2^k - c_1^i) = \max(2^k - c_1^i, c_1^i)$. Thus the set $T_1^1, T_1^2, \ldots, T_1^k$ is invariant under complementation of variables. Since complementation of the *function* has the same effect on c_1^i as complementing the term x_i, the set $\{T_1^i\}$ is also invariant under complementing the function. Finally, since $\{R_1^i\}$ is reordered according to decreasing size, a permutation of the variables has no effect on the set $\{T_1^i\}$.

Corollary.   The same reordered set $\{T_1^i\}$ which is invariant for the group G is of course invariant for the subgroup H. Here, however, no finer distinction is possible for H, since complementation of the function has effects on $\{T_1^i\}$ already obtainable by complementation of variables.

Definition.   The quantities $T_1^1, T_1^2, \ldots, T_1^k$ are called the *first order invariants* under G of the family for which they arise. (This set is also *the set of first order invariants under H*.)

Example.   Let $f(v, x, y, z)$ have the truth table shown in Figure VIII-2.

Fig. VIII-2.   The Boolean function $f(v, x, y, z) = v \oplus vx'z' \oplus xy'z \oplus v'x'yz$.

Then $c_0 = 7$, the total number of 1's in the table. Hence $T_0 = \max(7, 2^4 - 7) = 9$. Next, $R_1^v = \max(5, 2^4 - 5) = 11$, $R_1^x = \max(7, 2^4 - 7) = 9$, $R_1^y = \max(7, 2^4 - 7) = 9$, and $R_1^z = \max(5, 2^4 - 5) = 11$. Thus the set of first order invariants is $\{T_1^i\} = \{11, 11, 9, 9\}$, while the zero order invariant (under G) is $T_0 = 9$.

3. THE HIGHER ORDER INVARIANTS

Definition. For every pair (i, j) with $1 \leqq i < j \leqq k$, define $c_2^{ij} = \sum_{x_1, x_2, \ldots, x_k} (f(x_1, x_2, \ldots, x_k) \oplus x_i \oplus x_j)$. The complement of c_2^{ij} is $2^k - c_2^{ij}$. That ordering and complementing of variables and complementation of the function f, consistent with producing the $\{R_1^i\}$ in descending order, which makes the sequence of $\{c_2^{ij}\}$ numerically greatest, gives the second-order invariants $T_2^{1,2}, T_2^{1,3}, \ldots, T_2^{2,3}, \ldots, T_2^{k-1,k}$.

Notes:

1. The higher order invariants are defined in analogous fashion.

2. The invariants of *odd* order are the same with both G and H as the operator groups. The invariants of even order are in general different.

3. The reordering of the rth order invariants must be consistent with all the lower order invariants. It is only the first order invariants which may be ordered in descending fashion without regard to previous decisions.

4. The invariance of the higher order invariants under G obviously follows from the nature of the operators of G and the construction of the invariants.

Example. Consider the truth table in Figure VIII-3.

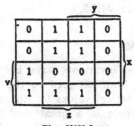

Fig. VIII-3.

Since
$$c_0 = 8, \ T_0 = 8 .$$

Next
$$c_1^z = 8, \ c_1^x = 10, \ c_1^y = 10, \ c_1^z = 4 .$$

Thus
$$R_1^v = 8, \ R_1^x = 10, \ R_1^y = 10, \ R_1^z = 12 ,$$

and
$$T_1^1 = 12, \ T_1^2 = 10, \ T_1^3 = 10, \ T_1^4 = 8 .$$

That is, $f(v, x, y, z)$ has been replaced by either $f(z', x, y, v)$, or $f(z', x, y, v')$, or $f(z', y, x, v)$, or $f(z', y, x, v')$.

Corresponding to these alternatives, one computes the values of

1. $c_2^{t'x}, c_2^{t'y}, c_2^{t'v}, c_2^{xy}, c_2^{xv}, c_2^{yv},$
2. $c_2^{t'x}, c_2^{t'y}, c_2^{t'v'}, c_2^{xy}, c_2^{xv'}, c_2^{yv'},$
3. $c_2^{t'y}, c_2^{t'x}, c_2^{t'v}, c_2^{yx}, c_2^{yv}, c_2^{xv},$
4. $c_2^{t'y}, c_2^{t'x}, c_2^{t'v'}, c_2^{yx}, c_2^{yv'}, c_2^{xv'}.$

These four sets of numbers are

$$\{10, \ 6, 12, 8, 6, 6\}, \qquad \{10, \ 6, 4, 8, 10, 10\},$$
$$\{6, 10, 12, 8, 6, 6\}, \quad \text{and} \quad \{6, 10, 4, 8, 10, 10\}.$$

Note that the *sum* of these six numbers has the same value (forty-eight) in all four cases. Next, choosing the lexicographically greatest assortment, $T_2^{12} = 10$, $T_2^{13} = 6$, $T_2^{14} = 12$, $T_2^{23} = 8$, $T_2^{24} = 6$, $T_2^{34} = 6$. The order of the variables is now firmly established as z', x, y, v. For the third-order invariants, it suffices to compute $c_3^{t'xy}, c_3^{t'xv}, c_3^{t'yv}$, and c_3^{yv}. These are $c_3^{t'xy} = 8$, $c_3^{t'xv} = 6$, $c_3^{t'yv} = 10$, $c_3^{xyv} = 8$. Hence $T_3^{123} = 8$, $T_3^{124} = 6$, $T_3^{134} = 10$, $T_3^{234} = 8$. Finally, the fourth order invariant is $T_4^{1234} = c_4^{t'xyv} = 8$. This completes the determination of the invariants for the logical function shown in Figure VIII-3. The symmetry $f(v, x, y, z) = f'(v, x', y', z')$ for this function is suggested by the resemblance among the four competing candidates for the second-order invariants.

This completes the determination of the invariants for the logical function shown in Figure VIII-3.

4. CANONICAL FORMS

By the time all the invariants have been determined, a certain "natural ordering" has been assigned to the variables, based on the criterion of highest numerical values for the successive orders of invariants. (Of course there is a great deal of arbitrariness in this "natural ordering." For example, a criterion of *lowest* numerical value for the invariants could have been adopted.) In the case that two variables occur symmetrically in $f(x_1, x_2, \ldots, x_k)$, the criterion of highest invariants will be unable to decide between them. Since they are symmetric, however, it is mathematically irrelevant which comes first. In the next section, it will be shown that no other ambiguities arise—that is, the full set of invariants completely characterizes the family of functions from which it arises.

Definition. Let $g(x_1, x_2, \ldots, x_k)$ be the result of permuting and complementing f itself, as determined by the computation of the invariants. Then $g(x_1, x_2, \ldots, x_k)$ is called the *canonical form* of $f(x_1, x_2, \ldots, x_k)$, and the truth table for g is the *canonical form* of the truth table for f.

Example. The canonical form for the truth table in Figure VIII-3 is shown in Figure VIII-4.

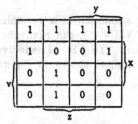

Fig. VIII-4. The Truth Table of Figure VIII-3 "reduced" to canonical form.

For a logical function which is already in canonical form, $c_0 = T_0$, and in general, $c_r^{ij\cdots m} = T_r^{ij\cdots m}$. This property can be used to define the canonical form. These equalities are readily verified for the truth table of Figure VIII-4.

5. THE RADEMACHER-WALSH EXPANSION COEFFICIENTS

Rademacher defined a set of functions $\{r_i(x)\}$ on the half-open interval $[0, 1)$, $i = 0, 1, 2, \ldots$, according to the rule:

$$r_i(x) = \begin{cases} 1 \text{ if } \dfrac{m}{2^i} \leq x < \dfrac{m+1}{2^i}, & m \text{ even} \\[2mm] -1 \text{ if } \dfrac{m}{2^i} \leq x < \dfrac{m+1}{2^i}, & m \text{ odd} . \end{cases}$$

Walsh proved that the collection of finite products of Rademacher functions form a *complete* orthonormal system of functions for the interval $[0, 1)$.

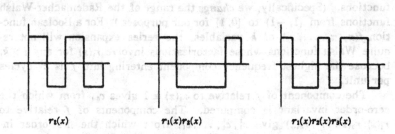

$r_2(x)$ $r_1(x)r_2(x)$ $r_1(x)r_2(x)r_3(x)$

Fig. VIII-5. Some typical Rademacher-Walsh functions.

Some typical Rademacher-Walsh functions are shown in Figure VIII-5. The completeness theorem implies that every real-valued function on [0, 1) with only finitely many discontinuities is a linear combination of these Rademacher-Walsh functions, and the linear combination is unique.

The truth table for a function of k variables, when written in the usual vertical way (Fig. VIII-6-A), can easily be converted into a function on [0, 1), as in Figure VIII-6-B. The individual variables can also be represented as functions, as shown in Figure VIII-7.

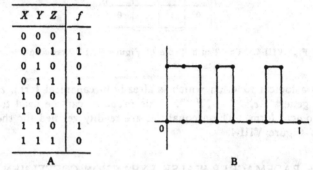

X	Y	Z	f
0	0	0	1
0	0	1	1
0	1	0	0
0	1	1	1
1	0	0	0
1	0	1	0
1	1	0	1
1	1	1	0

A   B

Fig. VIII-6.   Representation of a truth table as a function on the
interval [0, 1).

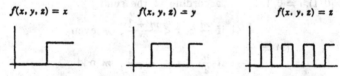

$f(x, y, z) = x$ $f(x, y, z) = y$ $f(x, y, z) = z$

Fig. VIII-7.   Graphical representation of the logical variables.

The graphs for the individual variables identify them as essentially Rademacher functions. Their modulo 2 sums are basically the Walsh functions. (Specifically, we *change* the range of the Rademacher-Walsh functions from {1, −1} to {0, 1} for our purposes.) For a Boolean function $f(x_1, x_2, \ldots, x_k)$ of k variables, the series expansion will not require Walsh functions whose factorizations involve $r_i(x)$ for any $i > k$, because the highest frequency component entering into f is 2^k cycles per unit.

The component of f relative to $r_0(x) \equiv 1$ gives c_0, from which the zero-order invariant is computed. The components of f relative to $r_1(x), r_2(x), \ldots, r_k(x)$ give $c_1^1, c_1^2, \ldots, c_1^k$, from which the first-order invariants of f are computed. The higher order invariants of f correspond to the coefficients of f relative to the Walsh functions.

Since the coefficients of any function relative to a complete orthonormal system completely characterize that function, the constants $c_r^{ij\cdots m}$ completely specify the corresponding Boolean function. The substitutions allowed in the determination of the invariants $T_r^{ij\cdots m}$ are precisely those allowed by the group G of symmetries. Hence

> *Theorem 3.* The 2^k invariants $T_r^{ij\cdots m}$, for $0 \leqq r \leqq k$, completely characterize the family of functions to which $f(x_1, x_2, \ldots, x_k)$ belongs, under G.

Note: The 2^k functions generated by $\{r_0(x), r_1(x), \ldots, r_k(x)\}$ are linearly independent (in fact, orthogonal) 2^k-dimensional vectors which therefore span the vector space of k-variable logical functions. The uniqueness of the expansion coefficients is thus algebraic as well as analytic. These same functions also appear in the guise of the Reed-Muller family of error correcting codes.

6. PRACTICAL METHODS OF COMPUTING THE INVARIANTS

The invariants $T_r^{ij\cdots m}$ are readily computed from the "proto-invariants" $c_r^{ij\cdots m}$. These in turn can be obtained directly from the truth table, and by rather simple techniques.

If the Boolean function is represented by its Karnaugh chart it is possible to make stencils such that the number N of 1's visible through

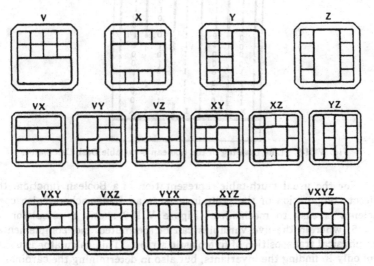

Fig. VIII-8. Stencils which yield proto-invariants for $k = 4$.

the stencil is $N_r^{ij\cdots m} = \tfrac{1}{2}(T_0 + c_r^{ij\cdots m} - 2^{k-1})$. Although the proto-in-variants $c_r^{ij\cdots m}$ are readily recoverable from the stencil-numbers $N_r^{ij\cdots m}$, it is frequently just as convenient to base the invariants on the $N_r^{ij\cdots m}$ directly.   The stencils for $k = 4$ are shown in Figure VIII-8.

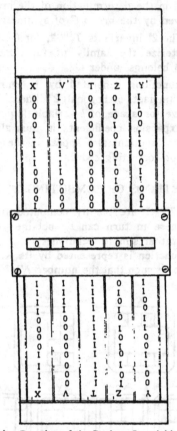

Fig. VIII-9.   Line Drawing of the Boolean 5-variable Desk Calculator.

For the usual truth-table representation of a Boolean function, the direct computation of the proto-invariants as Rademacher-Walsh coefficients is easy to mechanize. Figure VIII-9 shows a calculator for $k = 5$, wherein the five variables can be permuted and complemented by physical transposition of five plastic strips. This calculator is useful not only in finding the invariants, but also in determining the cardinality of classes of Boolean functions.

7. APPLICATION TO FINDING ASYMMETRICAL LOGICAL FUNCTIONS

If all the first-order invariants of a Boolean function $f(x_1, \ldots, x_k)$ are distinct, no permutation of variables leaves the function unaltered. If there is a variable whose complementation leaves the function unchanged, the corresponding first-order invariant must be 2^{k-1} (to satisfy $T_1^z = 2^k - T_1^z$). If all the first-order invariants are distinct, there is at most one such invariant (say T_1^z) such that $T_1^z = 2^{k-1}$. If there is no such invariant, the function f is changed by all $2^{k+1}k!$ operators in G. If there is an invariant $T_1^z = 2^{k-1}$, the only possible symmetries for f are $f(x', y, \ldots, z) = f(x, y, \ldots, z)$ or $f'(x', y, \ldots, z) = f(x, y, \ldots, z)$. In either case it is easy to determine by inspection whether or not f is altered by the entire transformation group. (A necessary condition for either of the symmetries mentioned to occur is that all the first order invariants be multiples of four, which already requires a large number of variables.)

For $k < 5$, it is impossible to construct a Boolean function with all its first order invariants distinct. For $k = 5$, there are several hundred such examples. In all cases, one of these invariants is 2^{k-1}. In no such case can degeneracy result upon replacing t by t', because the first order invariants cannot be *distinct* multiples of four, until $k \gg 5$.

With $k = 6$, it is possible to find Boolean functions with all the first-order invariants distinct and unequal to 2^{k-1}. These functions necessarily belong to maximum-size equivalence classes.

8. RELATED RESULTS AND PROBLEMS

The method used by Slepian [36] in determining the number of equivalence classes of Boolean functions is a direct application of the formula ([32], p. 131)

$$C = \frac{1}{N} \sum_{g \in G} I(g),$$ (1)

attributed to Pólya, for the number of equivalence classes C established in a set S under the operation of a group G of N elements. In (1), $I(g)$ is the number of set elements $s \in S$ for which $g(s) = s$. A modification of (1), which leads to the number of maximum-sized equivalence classes, for certain groups G, is

$$C^* = \frac{1}{N} \sum_{g \in G} I(g) \chi(g),$$ (2)

where $\chi(g)$ is a suitable primitive group character of G. For further development of this approach, see [13]. For the more general problem of classifying the switching functions with n inputs and k outputs, see [20], which also contains an excellent bibliography.

Another interesting question is whether or not a given set of numbers can be the set of invariants for some symmetry class of Boolean functions. This can be partially answered, in the form of constraints, relationships, and inequalities satisfied by the invariants. The analogous invariant theory for symmetry groups differing in certain respects from both G and H has also been developed. Finally, more general mappings than those from B^n into B (especially the case B^n into B^m) have also been examined.

An interesting constraint satisfied by the "reduced" invariants $\{N_r^{ij\cdots m}\}$ is based on the combinatorial "principle of inclusion and exclusion" (see [32], Chapter 3). This principle asserts that from a set of N objects, of which N_x have property x, N_y have property y, N_{xy} have both properties x and y, etc., the number of objects having *none* of the properties $x, y, \ldots$ is the alternating sum $N - N_x - N_y - \cdots + N_{xy} + \cdots - N_{xyz} - \cdots$. In terms of the reduced invariants, this leads to the constraint

$$N_0 - \sum N_1^i + \sum N_2^{ij} - \sum N_3^{ijk} + - \cdots = 2^{n-1} f(0, 0, \ldots, 0) , \qquad (3)$$

where f is the canonical member of the class whose invariants are the $\{N_r^{ij\cdots m}\}$, and n is the number of variables. For the function shown in Figure VIII-4.

$N_0 = 8$

$N_1^1 = 2, N_1^2 = 3, N_1^3 = 3, N_1^4 = 4,$

$N_2^{12} = 5, N_2^{13} = 3, N_2^{14} = 6, N_2^{23} = 4, N_2^{24} = 3, N_2^{24} = 3,$

$N_3^{123} = 4, N_3^{124} = 5, N_3^{134} = 3, N_3^{234} = 4,$

$N_4^{1234} = 4.$

The alternating sum is then: $8 - 12 + 24 - 16 + 4 = 8 = 2^{4-1} \cdot 1$.

The corresponding constraint for the *absolute* sum is

$$N_0 + \sum N_1^i + \sum N_2^{ij} + \sum N_3^{ijk} + \cdots = 2^{n-1}[N_0 + f(1, 1, \ldots, 1)] . \qquad (4)$$

This follows from the fact that each of the positions in the truth table is counted by *half* the 2^n invariants, except for the position $(1, 1, \ldots, 1)$, which is counted by all of them. For the example of Figure VIII-3, this sum is $8 + 12 + 24 + 16 + 4 = 64 = 8(N_0 + 0)$.

These constraints are easily transformed into statements about the invariants $\{T_r^{ij\cdots m}\}$. The goal here would be to find a *complete set* of constraints, so that given 2^n integers, it could easily be determined whether or not they were the set of invariants for some Boolean function of n variables.

REFERENCES

[1] Albert, A. A., *Fundamental Concepts of Higher Algebra*, 1st Edition, The University of Chicago Press, Chicago, 1956.

[2] Bruck, R., and H. Ryser, "The nonexistence of certain finite projective planes," Can. J. Math., Vol. 1, 1949, p. 88.

[3] de Bruijn, N. G., "A combinatorial problem," *Koninklijke Nederlands Akademie van Wetenschappen, Proceedings*, Vol. 49 (Part 2), 1946, pp. 758-764.

[4] Dickson, L. E., *Linear Groups*, Chapter IV, Leipzig, 1901; reprinted, Dover, New York, 1958.

[5] Feller, W., *An Introduction to Probability Theory and Its Applications*, Vol. 1, 1st Edition, John Wiley & Sons, Inc., New York, 1950.

[6] Ford, L. R., Jr., *A Cyclic Arrangement of m-tuples*, Report No. P-1071, Rand Corporation, Santa Monica, California, April 23, 1957.

[7] Franklin, J. N., "Deterministic simulation of random processes," *Mathematics of Computation*, Vol. 17, 1963, pp. 28-59.

[8] Franklin, J. N., "On the equidistribution of pseudo-random numbers," *Quart. Appl. Math.*, Vol. 16, 1958, pp. 183-188.

[9] Fricke, R., *Lehrbuch der Algebra*, Vol. III, Braunschweig, 1928, p. 50.

[10] Gauss, C. F., *Disquisitiones Arithmeticae*, Leipzig, 1801.

[11] Gilbert, E. N., *Quasi-Random Binary Sequences*, Bell Telephone Laboratories Memorandum MM-53-1400-42, November 27, 1953.

[12] Ginsburg, S. *An Introduction to Mathematical Machine Theory*, Addison-Wesley, Reading, Mass., 1962.

[13] Golomb, S. W., "A mathematical theory of discrete classification," in *Information Theory*, (Proceedings of the Fourth London Symposium on Information Theory), C. Cherry, ed., Butterworth & Co., 1961.

[14] Golomb, S. W., "Problem E 1319," *Am. Math. Monthly*, Vol. 66 (Part I) No. 64, January, 1959.

[15] Golomb, S. W., et al., *Digital Communications with Space Applications*, Chaps. I, II, IV, Prentice-Hall, Englewood Cliffs, N. J., 1964.

[16] Goncharov, V., "Du Domaine d'Analyse Combinatoire," *Bull. Acad. Sci. URSS, Ser. Math.*, Vol. 8, 1944, pp. 3-48.

[17] Good, I. J., "Normal recurring decimals," *J. London Math. Soc.*, Vol. 21 (Part 3), 1946, pp. 169-172.

[18] Hall, M., Jr., "A survey of combinatorial analysis," in Kaplansky, Hall, Hewitt, and Fortet, *Some Aspects of Analysis and Probability*, John Wiley & Sons, Inc., New York, 1958.

[19] Hall, M., Jr., "A survey of difference sets," *Proc. Am. Math. Soc.*, Vol. 7, 1956, pp. 975-986.

[20] Harrison, M. A., "On the number of classes of (n, k) switching networks," *J. Franklin Institute*, Vol. 276, No. 4, October, 1963, pp. 313-327.

[21] Kelly, J. B., "A characteristic property of quadratic residues," *Proc. Am. Math. Soc.*, Vol. 5, 1954, p. 38.

[22] Kummer, E. E., "Uber die Zerlegung der aus Wurzeln der Einheit gebildeten complexen Zahlen in ihre Primfactoren," *J. reine angew. Math.*, Vol. 35, 1847, pp. 327-367.

[23] Marsh, R. W., *Table of Irreducible Polynomials over GF* (2) *through Degree 19*, distributed by Office of Technical Services, Commerce Department, Washington, D. C., October 24, 1957.

[24] Moore, E. F., *Sequential Machines: Selected Papers*, Addison-Wesley, Reading, Mass., 1964.

[25] Nadler, M., and A. Sengupta, "Shift register code for indexing applications," *Communications of the ACM*, Vol. 2, No. 10, October, 1959, pp. 40-43.

[26] Ore, O., "Contributions to the theory of finite fields," *Transactions of the Am. Math. Soc.*, Vol. 36, 1934, pp. 243-274.

[27] Paley, R. E. A. C., "On orthogonal matrices," *J. Math. Phys.*, Vol. 12, 1933, pp. 311-320.

[28] Perron, O., "Bemerkungen über die Verteilung der Quadratischen Reste," *Math. Zeitschrift*, Vol. 56, 1952, pp. 122-130.

[29] Peterson, W. W., *Error-Correcting Codes*, John Wiley & Sons, Inc., New York, 1961.

[30] Rees, D., "Notes on a paper by I. J. Good," *J. London Math. Soc.*, Vol. 21 (Part 3), 1946, pp. 169-172.

[31] Reuschle, K. G., *Tafeln Complexer Primzahlen, welche aus Wurzeln der Einheit gebildet sind: Auf dem Grunde Kummerschen Theorie der complexen Zahlen*, Akademie der Wissenschaften, Berlin, 1875.

[32] Riordan, J., *An Introduction to Combinatorial Analysis*, John Wiley & Sons, Inc., New York, 1958.

[33] Ryser, H. J., *Combinatorial Mathematics*, The Carus Mathematical Monographs, Mathematical Association of America, John Wiley & Sons, Inc., 1963.

[34] Selmer, E., "On the irreducibility of certain trinomials," *Math. Scandinavica*, Vol. 4, 1956, pp. 287-302.

[35] Shannon, C. E., "Communication theory of secrecy systems," *Bell System Tech. J.*, Vol. 28, No. 4, October, 1949, pp. 656-715.

[36] Slepian, D., "On the number of symmetry types of Boolean functions of n variables," *Can. J. Math.*, Vol. 5, No. 2, pp. 185-193.

[37] Swan, R. G., "Factorization of polynomials over finite fields," *Pacific J. Math.*, Vol. 12, 1962, pp. 1099-1106.

[38] Taussky, O., and J. Todd, "Generation of pseudo-random numbers," *Symposium on Monte Carlo Methods*, H. A. Meyer, ed., John Wiley & Sons, Inc., New York, 1956, pp. 15-18.

[39] Van der Waerden, B. L., *Modern Algebra*, Vol. I (1949) and Vol. II (1950), Frederick Ungar Publishing Company, New York.

[40] Zierler, N., "Linear Recurring Sequences," *Linear Sequential Switching Circuits*, W. Kautz, ed., Holden-Day, Inc., San Francisco, 1965.

[41] Zierler, N., "On the theorem of Gleason and Marsh," *Proc. Am. Math. Soc.*, Vol. 9, 1958, pp. 236-237.

Chapter IX

SELECTIVE UPDATE

1. MERSENNE PRIMES

The Mersenne primes (cf. p. 37) are the prime numbers of the form $2^n - 1$. A necessary condition for $M_n = 2^n - 1$ to be prime is that n itself be prime. It is conjectured that there are infinitely many Mersenne primes. So far, forty-nine values of n have been shown to be "Mersenne exponents", that is, values of n such that M_n is prime. The first 43 of these are known to be the first 43 Mersenne exponents. The region above that has not been searched completely, so that there may be Mersenne exponents between the 43rd and the 49th in addition to the ones shown here.

2	31	1279	9941	110,503	2,976,221	30,402,457
3	61	2203	11,213	132,049	3,021,377	32,582,657
5	89	2281	19,937	216,091	6,972,593	37,156,667
7	107	3217	21,701	756,839	13,466,917	42,643,801
13	127	4253	23,209	859,433	20,996,011	43,112,609
17	521	4423	44,497	1,257,787	24,036,583	57,885,161
19	607	9689	86,243	1,398,269	25,964,951	74,207,281

2. THE $Z(n)$ CONJECTURE AND THE DE BRUIJN GRAPH

The conjecture (p. 174) that the maximum number of cycles into which B_n, the de Bruijn graph of order n, can be decomposed is

$$Z(n) = \frac{1}{n} \sum_{d|n} \phi(d) 2^{n/d}$$

was proved by Johannes Mykkeltveit of Norway [2], who built on a foundation laid by Harold M. Fredricksen [3] and Abraham Lempel [4]. In fact,

225

Mykkeltveit proves the stronger conjecture, due to Lempel, that the minimum number of vertices which, if removed from B_n, will leave a graph with no cycles, is $Z(n)$.

N. G. de Bruijn has discovered [5] a surprisingly early reference to both the problem (posed by A. de Riviere) and the solution (provided by C. Flye Sainte-Marie) of determining the number (namely $2^{2^{n-1}-n}$) of "de Bruijn sequences" of period 2^n. Although de Bruijn suggests that these sequences be renamed accordingly, as a practical matter this is most unlikely to occur. [6]

Another instance of the true antiquity of this subject is that the "prefer ONEs" construction (cf. p. 131) for a de Bruijn sequence of span n goes back to M. H. Martin [7] in 1934. A series of papers by Harold M. Fredricksen, beginning in 1970, and mostly appearing in the *Journal of Combinatorial Theory*, give practical constructions for fairly large subclasses of the set of de Bruijn sequences. (For these references, see the Comprehensive Bibliography.)

For an interesting new result, see "The directed genus of the de Bruijn graph" [27], by A.W. Hales and N. Hartsfield.

3. LENGTHS OF CYCLES FROM RANDOM PERMUTATIONS

In Chapter 7, the constant $\lambda = 0.6243299\ldots$ was introduced (see Theorem 6, p. 179) as the limit of the expected fraction of N objects which are on the longest cycle of a random permutation of those objects, as $N \to \infty$.

Several expressions for λ as a definite integral are known (see p. 192), but it has still not been determined whether λ is rational, irrational algebraic, or transcendental. However, λ has been found to arise in several additional contexts.

Paul Purdom and John Williams [8] found that λ is involved in the expected length of the longest cycle of a random "into" mapping of N objects: $\lim_{N \to \infty} E(M(F)) = \lambda \sqrt{\frac{\pi N}{2}}$.

(A permutation is an "onto" mapping of N objects.) This argument was condensed by Donald Knuth [9], who subsequently observed, with Luis Trabb Pardo [10], that λ is also identical to a constant discovered by Dickman [11], which is the limit of the average number of digits in the largest prime factor of an m-digit number, divided by m, as $m \to \infty$. (More generally, the expected length of the kth longest cycle of a random permutation on m objects corresponds to the average number of digits in the kth largest prime factor of an m-digit number, in the limit as $m \to \infty$. Following Knuth,

λ is now referred to as "Golomb's constant", or as "the Golomb-Dickman constant".

Also, λ arises when one considers the average degree of the highest-degree-factor among all polynomials of degree n over the finite field GF(q). Interesting discussions and generalization can be found in the article by Arratia, Barbour, and Tavaré [28], and by Lagarias [29], who further showed that λ occurs in an expression that also contains Euler's constant γ.

The extensive numerical calculation of λ carries out by William C. Mitchell [12] has been further extended to give $\lambda = 0.624329988543550\ldots$.

4. FACTORIZATION OF TRINOMIALS

On page 93, the following conjecture appears: "All irreducible factors of $F(x) = x^{2^n+1} + x^{2^{n-1}-1} + 1$ have degrees dividing $6(n-1)$." Mills and Zierler [13] have proved a slightly stronger result, namely that the degree of every irreducible factor of $F(x)$ divides either $2(n-1)$ or $3(n-1)$, but not $n-1$.

Tables V-1 and V-2 (pages 97–107), yielding information about irreducible and maximal (i.e., primitive) trinomials over $GF(2)$ through degree 45, have been significantly extended by Brillhart and Zierler in two papers, [14] and [15], in which they list all irreducible trinomials over $GF(2)$ of degree up to 1000, and give the periods of these irreducible trinomials for those degrees n for which the factorization of $2^n - 1$ is known.

A somewhat elementary proof of Richard Swan's theorem, from which it is known that all trinomials of degree $8n$ over $GF(2)$ are reducible (see page 96), and related results, can be found on pages 159–171 of Berlekamp's book [16], which also indicates applications of Swan's theorem to trinomials of degrees other than multiples of 8, thus substantially modifying the first sentence on page 108 of this book.

If the degree of a trinomial is a Mersenne exponent, then if it is irreducible it must be primitive, over GF(2). This has led to a world-wide distributed search, "The Great Trinomial Hunt" [30], Recently (2016)the search found three primitive trinomials of the huge degree $n = 74,207,281$, the largest (49th) Mersenne exponent currently known. One such has $a = 9,156,813$.

5. CROSS-CORRELATION OF PN SEQUENCES

On pp. 82–85, there is a largely empirical study of regularities in the cross-correlation of two PN sequences of the same period $p = 2^n - 1$. There has been an explosion of new results on this subject, based on improved techniques, in recent years. Some important earlier papers were those of Kasami [17], Gold [18], Golomb [19], Niho [20], and Helleseth [21]. Several of these papers observe the equivalence of the cross-correlation problem to the problem of *weight distribution,* or *weight enumeration,* for cyclic codes used in algebraic error correction. The principal methodology involves the use of *trace mappings* from $GF(2^n)$ to one of its subfields. If K is a finite algebraic extension field of F_2, and α is an element of K, then the "relative trace" of α from K to F, denoted by $Tr_{K/F}(\alpha)$, is the sum of all the conjugates of α in K over F, and is an element of F. Moreover, $Tr_{K/F}$ is a linear transformation from K to F, and, if F is regarded as the ground field, with K as a vector space over F, then $Tr_{K/F}$ is a linear functional K to F.

Some of the cross-correlation results obtained in this way are as follows: Define the cross-correlation function (taking $a_i = \pm 1$)

$$C_q(\tau) = \sum_{i=1}^{p} a_i a_{qi+\tau}$$

where $p = 2^n - 1$ is the period of the maximal linear shift register sequence $\{a_i\}$, of degree n. Then, provided that $(p, q) = 1$, $\{a_{qi}\}$ is also a maximal linear shift register sequence of degree n and period p. It is the same sequence as $\{a_i\}$ if and only if $q \in \{1, 2, 4, 8, \ldots, 2^{n-1}\}$, in which case $C_q(\tau)$ becomes the autocorrelation function, and takes on only the two values $C_q(0) = p$ and $C_q(\tau) = -1$ for all $\tau \not\equiv 0 \pmod{p}$. Assuming $(p, q) = 1$, if $q \notin \{1, 2, 4, 8, \ldots, 2^{n-1}\}$ then $C_q(\tau)$ must assume more than two values, and the case of greatest interest is when $C_q(\tau)$ takes on exactly three distinct values. Sufficient conditions for three-valued cross-correlation include the following.

a. For degree n, any $q = 2^k + 1$ such that n/e is odd, where $e = (k, n)$, leads to three-valued cross-correlation, where the three values which occur are: -1, $-1 + 2^{(n+e)}/2$, and $-1 - 2^{(n+e)}/2$.

b. With n, k and e as before, and n/e odd, $q = 2^{2k} - 2^k + 1$ leads to three-valued cross-correlation, where again the three values which occur are: -1, $-1 + 2^{(n+e)}/2$, $-1 - 2^{(n+e)}/2$.

c. For any q which leads to three-valued cross-correlation, all other members of its cyclotomic coset, namely the values $2^s q \pmod{p}$ for $1 \leq s \leq n$, lead to three-valued cross-correlation with the same three values. So too do the members of the inverse cyclotomic coset, i.e. the coset containing q' such that $qq' \equiv 1 \pmod{p}$, where $p = 2^n - 1$.

d. The choice $q = 2^{n-2} + 3$ leads to three-level cross-correlation for all odd $n \leq 17$. This is conjectured to hold for *all* odd n.

These results on cross-correlation of linear sequences have been applied to the problem of finding *sets* of signals such that the cross-correlation between any two members of the set is low. The Gold codes [22] are an important early example of this phenomenon. Using the calculus of variations, L. R. Welch made a fundamental contribution [23] to this subject by obtaining a *lower bound* to the maximum value of cross-correlation which must occur between pairs of signals from a signal set. Recent work has focused on trying to find sets of signals, binary or otherwise, which attain the Welch bound. For this purpose, there are "Kasami sequences" derived from Kasami's original paper [17], "bent sequences" based on the work of Rothaus [24], and the "group character sequences" of Scholtz and Welch [25]. See also the excellent survey article by Sarwate and Pursley [26].

There is considerable practical interest in the behavior of both the auto-correlation and the cross-correlation functions of maximum length linear shift register sequences when these correlations are computed over only a part of the period. A search of the literature (cf. the Comprehensive Bibliography) reveals quite a few papers which have addressed this subject, but the results to date are mostly ad hoc, sporadic, or only narrowly applicable. Since these problems can be reformulated in terms of weight distribution and weight enumeration of group codes, there is some hope that major progress may yet occur on this front.

Two important recent additions to the literature are "Binary m-Sequences with Three-Valued Cross-correlation: A Proof of Welch's Conjecture," by Canteaut, Charpin and Dobertin, [31], and "Open Problems on the Cross-Correlation of m-Sequences", by Tor Helleseth [32], which mostly concerns m-sequences over finite fields larger than GF(2).

6. ADDITIONAL REFERENCES

There is an extensive section on shift register sequences in the *Handbook of Finite Fields* [33], by Mullen and Panario.

Other useful references are *Introduction to Finite Fields and their Applications* [34] by Lidl and Niederreiter and *Signal Design for Good Correlation*, [35] by Golomb and Gong. A valuable resource for information on Mersenne exponents, primitive trinomials over GF [2], the Golomb-Dickman constant, etc, is Neil Sloane's "Online Encyclopedia of Integer Sequences" [36].

REFERENCES FOR CHAPTER IX

[1] David Slowinski, Searching for the twenty-seventh Mersenne prime, *Journal of Recreational Mathematics*, vol. 11, no. 4, 1978–79, pp. 258–261.

[2] Johannes Mykkeltveit, A proof of Golomb's Conjecture for the de Bruijn
     graph, *Journal of Combinatorial Theory (B)*, vol. 13, no. 1, August, 1972,
     pp. 40–45.

[3] Harold M. Fredricksen, *Disjoint Cycles from the de Bruijn Graph*, Ph.D.
     Thesis, Department of Electrical Engineering, University of Southern Cali-
     fornia, Los Angeles, California, June, 1968.

[4] Abraham Lempel, On the extremal factors of the de Bruijn graph, *Journal
     of Combinatorial Theory (B)*, vol. 11, no. 1, August, 1971, pp. 17–27.

[5] N. G. de Bruijn, Acknowledgement of priority to C. Flye Sainte-Marie on
     the counting of 2^n zeros and ones that show each n-letter word exactly once,
     Technical Report TH 75-WSK-06, Technische Hogeschool Eindhoven, 1975.

[6] A. de Rivière, Question no. 48, *L'Intermédiare des Mathématiciens*, vol. 1,
     no. 2, Feb., 1894, pp. 19–20, and C. Flye Sainte-Marie, Solution to ques-
     tion no. 48, *L'Intermédiaire des Mathématiciens*, vol. 1, no. 6, June, 1894,
     pp. 107–110.

[7] M. H. Martin, A problem in arrangements, *Bulletin of the American Math-
     ematical Society*, vol. 40, no. 12, December, 1934, pp. 859–864.

[8] Paul W. Purdom and John H. Williams, Cycle length in a random func-
     tion, *Transactions of the American Mathematical Society*, vol. 133, no. 2,
     September, 1968, pp. 547–551.

[9] Donald E. Knuth, *The Art of Computer Programming, vol. 2, Seminumeri-
     cal Algorithms*, Addison-Wesley Publishing Co., Reading, Mass., 1969, p. 8,
     Exercise 13, and p. 454.

[10] Donald E. Knuth and Luis Trabb-Pardo, Analysis of a simple factoriza-
     tion algorithm, *Theoretical Computer Science*, vol. 3, no. 3, December, 1976,
     pp. 321–348.

[11] Karl Dickman, On the frequency of numbers containing prime factors of
     a certain relative magnitude, *Arkiv for Matematik, Astronomi och Fysik*,
     vol. 22A, no. 10, Häfte 2, 1930, pp. 1–14.

[12] William C. Mitchell, An evaluation of Golomb's constant, *Mathematics of
     Computation*, vol. 22, no. 102, April, 1968, pp. 411–415.

[13] W. H. Mills and Neal Zierler, On a conjecture of Golomb, *Pacific Journal of
     Mathematics*, vol. 28, no. 3, March, 1969, pp. 635–640.

[14] Neal Zierler and John Brillhart, On primitive trinomials (mod 2), *Informa-
     tion and Control*, vol. 13, no. 6, December, 1968, pp. 541–554.

[15] Neal Zierler and John Brillhart, On primitive trinomials (mod 2), II, *Infor-
     mation and Control*, vol. 14, no. 6, June, 1969, pp. 566–569.

[16] Elwyn R. Berlekamp, *Algebraic Coding Theory*, McGraw-Hill Book Com-
     pany, New York, 1968.

[17] Tadao Kasami, *Weight Distribution Formula for Some Class of Cyclic Codes*,
     Report R-285, Coordinated Science Lab., University of Illinois, Urbana, Illi-
     nois, April, 1966.

[18] Robert Gold, Maximal recursive sequences with three-valued recursive cross-
     correlation function, *IEEE Transactions on Information Theory*, vol. IT-14,
     no. 1, January, 1968, pp. 154–156.

[19] Solomon W. Golomb, Theory of transformation groups of polynomials over
     $GF(2)$ with applications to linear shift register sequences, *Information Sci-
     ences*, vol. 1, no. 1, December, 1968, pp. 87–109.

[20] Yoji Niho, *Multi-valued cross-correlation functions between two maximal linear recursive sequences*, Ph.D. Thesis, Department of Electrical Engineering, University of Southern California, Los Angeles, California, June, 1972.

[21] Tor Helleseth, Some results about the cross-correlation function between two maximal linear sequences, *Discrete Mathematics*, vol. 16, no. 3, November, 1976, pp. 209–232.

[22] Robert Gold, Optimal binary sequences for spread spectrum multiplexing, *IEEE Transactions on Information Theory*, vol. IT–13, no. 4, October, 1967, pp. 619–621.

[23] Lloyd R. Welch, Lower bounds on the maximum cross-correlation of signals, *IEEE Transactions on Information Theory*, vol. IT-20, no. 3, May, 1974, pp. 397–399.

[24] Oscar S. Rothaus, On bent functions, *Journal of Combinatorial Theory (A)*, vol. 20, no. 3, May, 1976, pp. 300–305.

[25] Robert A. Scholtz and Lloyd R. Welch, Group characters: sequences with good correlation properties, *IEEE Transactions on Information Theory*, vol. IT-24, no. 5, September, 1978, pp. 537–545.

[26] D. V. Sarwate and M.B. Pursley, Cross-correlation properties of pseudorandom and related sequences, *Proceedings of the IEEE*, vol. 68, no. 5, May, 1980, pp. 593–619.

[27] "The directed genus of the de Bruijn graph", by Alfred W. Hales and Nora Hartsfield, *Discrete Math* 309 (17), September 2009, pp. 5259–5963.

[28] "On random polynomials over finite field". by Richard Arratia, A.D. Barbour and Simon Tavaré, *Math. Proc. Cambridge Phil. Soc.* (1993), 114, pp. 347–368.

[29] "Euler's constant: Euler's work and modern developments", by Jeffrey C. Lagarias, *Bulletin (new series) of the American Math. Soc.* volume 50, no. 4, October, 2013, pp. 527–628.

[30] "The great trinomial hunt", by Richard P. Brent and Paul Zimmermann, *AMS Notices* 58(2), February, 2011, pp. 233–239.

[31] "Binary *m*-Sequences with Three-Valued Cross-correlation: A Proof of Welch's Conjecture", by Anne Canteaut, Pascal Charpin, and Hans Dobbertin, *IEEE TRANS, INFO. THEORY*, vol. 46, no. 1, January, 2000, pp. 4–8.

[32] "Open Problems on the Cross-Correlation of *m*-Sequences," by Tor Helleseth, *Open Problems in Mathematical and Computational Science*, Ç.K. Koç (ed), Springer, 2014, pp. 163–179.

[33] *Handbook of Finite Fields*, by Gary L. Mullen and Daniel Panario, CRC Pres, 2013.

[34] *Introduction to Finite Fields and their Applications*, Second Edition, Rudolph Lidl and Harald Niederreiter, Cambridge University Press, 2008.

[35] *Signal Design for Good Correlation*, by Solomon Golomb and Guang Gong, Cambridge University Press, 2005.

[36] "Online Encyclopedia of Integer Sequences", maintained online by Neil J.A. Sloane, and regularly updated.

COMPREHENSIVE BIBLIOGRAPHY

References with an AD Number are available from the DDC (Defense Dcoumentation Center, Cameron Station, Alexandria, Virginia, 22314).

[1]   Ackroyd, M. H., "Synthesis of Efficient Huffman Sequences," *IEEE Transactions on Aerospace and Electronic Systems*, Vol. AES-8, 1972, pp. 2-6.

[2]   Ackroyd, M. H., "Huffman Sequences with Approximately Uniform Envelopes or Cross-Correlation Functions," *IEEE Transactions on Information Theory*, Vol. IT-23, 1977, pp. 620-623.

[3]   Aein, J. M., and J. W. Schwartz, Eds. "Multiple-Access to a Communication Satellite with a Hard-Limiting Repeater — Volume II: Proceedings of the IDA Multiple Access Summer Study," Inst. Defense Analyses, Rep. R-108, (AD 465789), 1965.

*[4] Albert, A. A., *Fundamental Concepts of Higher Algebra*, 1st Edition, University of Chicago Press, Chicago, 1956.

[5]   Alekseev, A. I., et al., *Teoriya I Premenenie Psevdosluchainykh Signalov, Nauka Publishers*, 1969.

[6]   Alltop, W. O., "Complex Sequences with Low Periodic Correlations," *IEEE Transactions on Information Theory*, Vol. IT-26, 1980, pp. 350-354.

[7]   Anderson, D. R., "Families of Equal-Length Shift Register Sequences Obtainable from a New Class of Cyclic Codes," *IEEE Int. Com. Conf. Digest*, No. 19C30, 1966, p. 179.

[0]   Anderson, D R , "A New Class of Cyclic Codes," *SIAM J. Appl. Math.*, Vol. 16, 1968, pp. 181-197.

[9]   Anderson, D. R., and P. A. Wintz, "Analysis of a Spread-Spectrum Multiple-Access System with a Hard Limiter," *IEEE Transactions on Communication Technology*, Vol. COM-17, 1969, pp. 285-290.

[10] Artom, A., "Choice of Prefix in Self-Synchronizing Codes," *IEEE Transactions on Communications*, April, 1972, pp. 253-254.

[11] Bailey, J. S., "Generalized Single-Ended Counters," *Journal of the Assoc. for Computing Machinery*, Vol. 13, No. 3, July, 1966, pp. 412-418.

[12] Bailey, J. S., and G. Epstein, "Single Function Shifting Counters," *Journal of the Assoc. for Computing Machinery*, Vol. 9, No. 3, July, 1962, pp. 375-378.

[13] Barker, R. H., "Group Synchronizing of Binary Digital Systems," *Communication Theory*, Butterworth, London, England, 1953.

[14] Bartee, T. C., and D. I. Schneider, "Computation with Finite Fields," *Information and Control*, Vol. 6, 1963, pp. 79-98.

[15] Bartee, T. C., and P. E. Wood, *Coding for Tracking Radar Ranging*, Tech. Rep. 318, MIT Lincoln Lab., Lexington, Mass., 1963.

[16] Baumert, L. D., *Lecture Notes in Mathematics — Cyclic Difference Sets*, Springer-Verlag, 1971.

[17] Baumert, L. D., and R. J. McEliece, "Weights of Irreducible Cyclic Codes," *Information and Control*, Vol. 20, 1972, pp. 158-175.

[18] Beard, J. T. B., Jr., and K. I. West, "Some Primitive Polynomials of the Third Kind," *Math. Comp.*, Vol. 28, No. 128, October, 1974, pp. 1166-1167 + microfiche.

[19] Bekir, N. E., R. A. Scholtz, and L. R. Welch, "Partial-Period Correlation Properties of PN Sequences," *1978 National Telecommunications Conf. Rec.*, Vol. 3, 1978, pp. 35.1.1-35.1.4.

[20] Berge, C., *The Theory of Graphs and Its Applications*, Methuen, London, 1962.

[21] Berge, C., *Principles of Combinatorics*, Academic Press, 1971.

[22] Berlekamp, E. R., "Factoring Polynomials Over Finite Fields," *Bell System Tech. J.* 46, 1967, pp. 1855-1859.

**[23] Berlekamp, E. R., *Algebraic Coding Theory*, McGraw-Hill, New York, 1968.

[24] Berlekamp, E. R., "Factoring Polynomials Over Large Finite Fields," *Math. Comp.* 24, (111), 1970, pp. 713-735.

[25] Birdsall, T., R. Heitmeyer, and K. Metzger, *Modulation by Linear Maximal Shift Register Sequences: Amplitude, Biphase and Complement-Phase Modulation*, Michigan Univ., Ann Arbor, Cooley Electronics Lab., December, 1971.

234

[26] Blachman, Nelson M., *General Criteria of ECM Effectiveness (U)*, Sylvania Elec. Sys. Report, pp. 35-115.

[27] Bloom, G. S., and S. W. Golomb, "Applications of Numbered Undirected Graphs," *Proceedings of the IEEE*, Vol. 65, 1977, pp. 562-570.

[28] Boehmer, A. M., "Binary Pulse Compression Codes," *IEEE Transactions on Information Theory*, Vol. IT-13, 1967, pp. 156-167.

[29] Borth, D. E., and M. B. Pursley, "Analysis of Direct Sequence Spread-Spectrum Multiple-Access Communication over Rician Fading Channels," *IEEE Transactions on Communications*, Vol. COM-27, 1979, pp. 1566-1577.

[30] Braasch, R. H., *Report SC-DR-65-406*, Sandia Corporation, 1965.

[31] Briggs, P. A. N., and K. R. Godfrey, "Pseudorandom Signals for the Dynamic Analysis of Multivariable Systems," *Proc. Inst. Elec. Eng.*, Vol. 113, 1966, pp. 1259-1267.

[32] Briggs, P. A. N., and K. R. Godfrey, "Autocorrelation Function of a 4-Level m-Sequence," *Electronics Letters*, Vol. 4, 1968, pp. 232-233.

[33] Briggs, P. A. N., and K. R. Godfrey, "New Class of Pseudorandom Ternary Sequences," *Electronics Letters*, Vol. 4, 1968, pp. 438-439.

[34] Briggs, P. A. N., and K. R. Godfrey, "Design of Uncorrelated Signals," *Electronics Letters*, Vol. 12, 1976, pp. 555-556.

[35] Briggs, P. A. N., P. H. Hammond, M. T. G. Hughes, and G. O. Plumb, "Correlation Analysis of Process Dynamics Using Pseudorandom Binary Test Perturbation," *Proc. Inst. Mech. Eng.*, Vol. 179, Pt. 3H, 1965, 37-51.

*[36] Bruck, R., and H. Ryser, "The Nonexistence of Certain Finite Projective Planes," *Can. J. Math.*, Vol. 1, 1949, p. 88.

*[37] de Bruijn, N. G., "A Combinatorial Problem," *Proc. Kon. Ned. Akad. Wet.*, Vol. 49 (Part 2), 1946, pp. 758-764.

[38] de Bruijn, N. G., "On A Function Occuring in the Theory of Primes," *J. Indian Math. Soc.*, A 15, 1951, pp. 25-32.

[39] de Bruijn, N. G., "On the Number of Positive Integers $\leq x$ and Free of Prime Factors $> y$," *Proc. Kon. Ned. Akad. Wet.*, A 54 (Indag. Math. 13), 1951, pp. 50-60.

**[40] de Bruijn, N. G., Acknowledgement of Priority to C. Flye Sainte-Marie on the Counting of 2^n Zeros and Ones that Show Each n-Letter Word Exactly Once, *Technical Report TH 75-WSK-06*, Technische Hogeschool Eindhoven, 1975.

[41] Bryant, P. R., R. G. Heath, and R. D. Killick, "Counting with Feedback Shift Registers by Means of a Jump Technique," *IRE Transactions on Electronic Computers*, EC-11, April, 1962, pp. 285-286.

[42] Bryant, P. R., and R. D. Killick, "Non-Linear Feed-back Shift Registers," *IRE Transactions on Computers*, June, 1962.

[43] Busacker, R. G., and T. L. Saaty, *Finite Graphs and Networks: An Introduction with Applications*, McGraw-Hill, New York, 1965.

[44] Calabi, L., and W. E. Hartnett, "A Family of Codes for the Correction of Substitution and Synchronization Errors," *IEEE Transactions on Information Theory*, Vol. IT-15, January, 1969, pp. 102-106.

[45] Calabro, D., and J. Paolillo, "Synthesis of Cyclically Orthogonal Binary Sequences of the Same Least Period," *IEEE Transactions on Information Theory*, Vol. IT-14, 1968, pp. 756-758.

[46] Caprio, J. R., "Strictly Complex Impulse-Equivalent Codes and Subsets with Very Uniform Distribution," *IEEE Transactions on Information Theory*, Vol. IT-15, 1969, pp. 695-706.

[47] Chakrabarti, N. B., and M. Tomlinson, "Design of Sequences with Specified Autocorrelation and Cross-Correlation," *IEEE Transactions on Communications*, Vol. COM-24, 1976, pp. 1246-1252.

[48] Chang, H. Y. P., *On Quasi-Linear Sequential Machines*, Dissertation No. 65-3191, Ph.D. Thesis, University of Illinois, Champagne, Illinois, 1964.

[49]   Christensen, J., "Enumerating Shift Register Sequences," Ph.D. Thesis, Department of E.E., University of Waterloo, June, 1974.

[50]   Chu, D. C., "Polyphase Codes with Good Periodic Correlation Properties," IEEE Transactions on Information Theory, Vol. IT-18, 1972, pp. 531-532.

[51]   Cohn, M., and A. Lempel, "On Fast m-Sequence Transforms," IEEE Transactions on Information Theory, Vol. IT-23, 1977, pp. 135-137.

[52]   Colley, L. E., Computation of Time-Phase Displacements of Binary Linear Sequence Generators, IEEE Computer Group Repository Paper, R-67-4, 1967.

[53]   Cook, C. E., and M. Bernfeld, Radar Signals, Academic Press, New York, 1967, Ch. 8.

[54]   Cooper, G. R., and R. W. Nettleton, "A Spread-Spectrum Technique for High-Capacity Mobile Communications," IEEE Transactions on Veh. Technology, Vol. VT-27, 1978, pp. 264-275.

[55]   Coven, E. M., and G. A. Hedlund, "Periods of Some Nonlinear Shift Registers," Journal of Combinatorial Theory, Series A, Vol. 27, No. 2, September, 1979, pp. 186-197.

[56]   Coven, E. M., G. A. Hedlund, and F. Rhodes, "The Commuting Block Maps Problem," Transactions of the American Mathematical Society, Vol. 249, No. 1, April, 1979, pp. 113-138.

[57]   Davis, W. A., "Generation of Delayed Replicas of Maximal Length Linear Binary Sequences," Proceedings of the IEEE, Vol. 113, February, 1966, pp. 295-296.

[58]   Davis, W. A., "Automatic Delay Changing Facility for Delayed m-Sequences," Proceedings of the IEEE, Vol. 54, June, 1966, pp. 913-914.

[59]   Dawson, R., et al., "Exact Markov Probabilities from Oriented Linear Graphs," Annals Math. Statistics, Vol. 28, 1957, pp. 946-956.

[60]   Delsarte, P., J. M. Goethals, and F. J. MacWilliams, "On Generalized Reed-Muller Codes and Their Relatives," Information and Control, Vol. 16, July, 1970, pp. 403-442.

[61]   Devyatkov, V. V., "The Realization of a Finite Automaton Using Two Shift Registers," Izvestia Nauk SSSR, Tekh. Kiber., No. 3, 1969, pp. 88-89.

**[62] Dickman, Karl, "On the Frequency of Numbers Containing Prime Factors of a Certain Relative Magnitude," Arkiv för Matematik, Astronomi och Fysik, Vol. 22A., No. 10, Häfte 2, 1930, pp. 1-14.

*[63]   Dickson, L. E., Linear Groups, Chapter IV, Leipzig, 1901; reprinted, Dover, New York, 1958.

[64]   Dowling, T. A., and R. J. McEliece, "Cross-Correlation of Reverse Maximal-Length Shift Register Sequences," JPL Space Programs Summary 37-53, Vol. 3, 1968, pp. 192-193.

[65]   Duvall, Paul F., "Decimation of Periodic Sequences," SIAM J. Appl. Math., Vol. 21, No. 3, November, 1971, pp. 367-372.

[66]   Duvall, P. F., Jr., and R. E., Kibler, "On the Parity of the Frequency of Cycle Lengths of Shift Register Sequences," J. Comb. Theory (A), Vol. 18, No. 3, May, 1975, pp. 357-361.

[67]   Eldert, C., H. J. Gray, Jr., H. M. Gurk, and M. Rubinoff, "Shifting Counters," AIEE Trans. (Commun. Electron.), Vol. 77, March, 1958, pp. 70-74.

[68]   Elspas, B., "Theory of Autonomous Linear Sequential Networks," IRE Transactions on Circuit Theory, Vol. CT-6, March, 1959, pp. 45-60.

[69]   Epstein, G., Non-Linear Single Function Shifting Registers, ITT Gilfillan, Inc., Van Nuys, California.

*[70]   Feller, W., An Introduction to Probability Theory and Its Applications, Vol. 1, 1st Edition, John Wiley & Sons, Inc., New York, 1950.

[71]   Ferguson, J. D., Some Properties of Mappings on Sequence Spaces, Ph.D. Dissertation, Yale University, 1962.

*[72]   Ford, L. R., Jr., A Cyclic Arrangement of m-tuples, Report No. P-1071, RAND Corporation, Santa Monica, California, April, 1957.

[73]   Forney, G. D., "Coding and its Application in Space Communications," IEEE Spectrum, Vol. 7, June, 1970, pp. 47-58.

[74]   Frank, R. L., "Polyphase Codes with Good Non-Periodic Correlation Properties," *IEEE Transactions on Information Theory*, Vol. IT-9, 1963, pp. 43-45.

[75]   Frank, F. L., and S. Zadoff, "Phase Shift Pulse Codes with Good Periodic Correlation Properties," *IRE Transactions on Information Theory*, Vol. IT-8, 1962, pp. 381-382.

*[76]   Franklin, J. N., "On the Equidistribution of Pseudo-Random Numbers," *Quart. Appl. Math.*, Vol. 16, 1958, pp. 183-188.

*[77]   Franklin, J. N., "Deterministic Simulation of Random Processes," *Mathematics of Computation*, Vol. 17, 1963, pp. 28-59.

[78]   Franklin, J. N., and S. W. Golomb, "A Function-Theoretic Approach to the Study of Nonlinear Recurring Sequences," *Pacific Journal of Mathematics*, Vol. 56, No. 2, 1975, pp. 455-468.

**[79] Fredricksen, H. M., *Disjoint Cycles from the de Bruijn Graph*, Ph.D. Thesis, Department of Electrical Engineering, University of Southern California, Los Angeles, California, June, 1968.

[80]   Fredricksen, H., "The Lexicographically Least de Bruijn Cycle," *J. Combinatorial Theory*, Vol. 9, No. 1, 1970, pp. 1-5.

[81]   Fredricksen, H., "Asymptotic Behavior of Golomb's $Z(n)$ Conjecture," *IEEE Transactions on Information Theory*, Vol. IT-16, July, 1970, p. 500.

[82]   Fredricksen, H., "Generation of the Ford Sequence of Length 2^n, n Large," *J. Combinatorial Theory (A)*, Vol. 12, No. 1, 1972, pp. 153-154.

[83]   Fredricksen, H., "A Class of Nonlinear de Bruijn Cycles," *J. Combinatorial Theory (A)*, Vol. 19, No. 2, 1975, pp. 192-199.

[84]   Fredricksen, H., "A Survey of de Bruijn Sequence Algorithms," *IDA-CRD Expository Report No. 16*, 1975.

[85]   Fredricksen, H., and I. Kessler, "Lexicographic Compositions and de Bruijn Sequences," *J. Combinatorial Theory (A)*, Vol. 22, No. 1, 1977, pp. 17-30.

[86]   Fredericksen, H., and J. Maiorana, "Necklaces of Beads in k Colors and k-ary de Bruijn Sequences," *Discrete Mathematics*, Vol. 23, No. 3, 1978, pp. 207-210.

[87]   Fredricsson, S., "Pseudo-Randomness Properties of Binary Shift Register Sequences," *IEEE Transactions on Information Theory*, Vol. IT-21, 1975, pp. 115-120.

*[88]   Fricke, R., *Lehrbuch der Algebra*, Vol. III, Braunschweig, 1928.

[89]   Gagliardi, R. M., "Rapid Acquisition Signal Design in a Multiple-Access Environment," *IEEE Transactions on Aerospace Electronic Systems*, Vol. AES-10, 1974, pp. 359-363.

[90]   Gallager, R., "Sequential Decoding for Binary Channels with Noise and Synchronization Errors," *MIT-RLE Report, AD266-879*, 1961.

[91]   Gaugg, Alfred, Manfred Seidel, and Hans Weinrichter "Grenzen der Eignung von Pseudozufallssignalen für Statistische Tests," *Arch. Elektron. Übertrag*, Vol. 27, No. 1, January, 1973, pp. 30-36.

*[92]   Gauss, C. F., *Disquisitiones Arithmeticae*, Leipzig, 1801. English translation by Arthur A. Clarke, S. J., Yale University Press, 1966.

[93]   Geffe, P. R., *Secure Electronic Cryptography*, Allerton Conference on Circuit and Systems Theory, 6th Proc., October, 1968, pp. 181-188.

*[94]   Gilbert, E. N., *Quasi-Random Binary Sequences*, Bell Telephone Laboratories Memorandum MM-53-1400-42, November 27, 1953.

[95]   Gilbert, E. N., "Synchronization of Binary Messages," *IRE Transactions on Information Theory*, September, 1960, pp. 470-477.

*[96]   Ginsburg, S., *An Introduction to Mathematical Machine Theory*, Addison-Wesley, Reading, Mass., 1962.

[97]   Gleason, A., et al., *A Method for Generating Irreducible Polynomials*, NSA Section 314 Report, May 26, 1955.

[98]   Godfrey, K. R., "Three-level m-Sequences," *Electronics Letters*, Vol. 2, 1966, pp. 241-243.

[99]   Godfrey, K. R., "The Theory of the Correlation Method of Dynamic Analysis and its Application to Industrial Processes and Nuclear Power Plants," *Measurement and Control*, Vol. 2, 1969, pp. T65-T72.

[100]   Goka, T., "An Operator on Binary Sequences," *SIAM Rev.*, Vol. 12, 1970, pp. 264-266.

[101]   Golay, M. J. E., "Complementary Series," *IRE Transactions on Information Theory*, Vol. IT-7, 1961, pp. 82-87.

[102]   Golay, M. J. E., "A Class of Finite Binary Sequences with Alternate Autocorrelation Values Equal to Zero," *IEEE Transactions on Information Theory*, Vol. IT-18, 1972, pp. 449-450.

[103]   Golay, M. J. E., "Notes on Impulse Equivalent Pulse Trains," *IEEE Transactions on Information Theory*, Vol. IT-21, 1975, pp. 718-721.

[104]   Gold, R., "Characteristic Linear Sequences and their Coset Functions," *SIAM J. Appl. Math.*, Vol. 14, 1966, pp. 980-985.

[105]   Gold, R., "Study of Correlation Properties of Binary Sequences," AF Avionics Lab., Wright-Patterson AFB, Ohio, Tech. Rep. AFAL-TR-66-234, (AD48858), 1966.

[106]   Gold, R., "Optimal Binary Sequences for Spread Spectrum Multiplexing," *IEEE Transactions on Information Theory*, Vol. IT-13, 1967, pp. 619-621.

[107]   Gold, R., "Study of Correlation Properties of Binary Sequences," AF Avionics Lab., Wright-Patterson AFB, Ohio, Tech. Rep. AFAL-TR-67-311, (AD826367), 1967.

[108]   Gold, R., "Maximal Recursive Sequences with 3-Valued Recursive Cross-Correlation Functions," *IEEE Transactions on Information Theory*, Vol. IT-14, 1968, pp. 154-156.

[109]   Gold, R., "Study of Multistate PN Sequences and Their Application to Communication Systems," Rockwell International Corp. Rep., (AD A025137), 1976.

[110]   Gold, R., and E. Kopitzke, "Study of Correlation Properties of Binary Sequences," Interim Tech. Rep. 1, Vols. 1-4, Magnavox Res. Lab., Torrance, California, (AD 470696-9), 1965.

[111]   Golomb, S. W., *Sequences with Randomness Properties*, The Glenn L. Martin Company, Baltimore, Md., 1955.

[112]   Golomb, S. W., "Problem E 1319," *Am. Math. Monthly*, Vol. 66, No. 1, January, 1959, p. 64.

[113]   Golomb, S. W., "A Mathematical Theory of Discrete Classification," *Information Theory*, (Proceedings of the Fourth London Symposium on Information Theory), C. Cherry, Ed., Butterworth & Co., 1961.

[114]   Golomb, S. W., "On Certain Nonlinear Recurring Sequences," *Am. Math. Monthly*, Vol. 70, No. 4, April, 1963.

[115]   Golomb, S. W., ed., *Digital Communications with Space Applications*, Prentice Hall, Englewood Cliff, N.J., 1964. Second edition, Peninsula Publishing, Los Altos, CA 1982.

[116]   Golomb, S.W. *Shift Register Sequences*, Holden-Day, San Francisco, CA 1967. Revised edition, Aegean Park Press, Laguna Hills, Ca 1982.

[117]   Golomb, S. W., "Irreducible Polynomials, Synchronization Codes, Primitive Necklaces, and the Cyclotomic Alegbra," *Combinatorial Mathematics and Its Applications*, Chapter 21, Proc. of the 1st U.N.C. Conf. On Combinatorial Math., April, 1967, University of North Carolina Press, 1969, pp. 358-370.

[118]   Golomb, S. W., "Theory of Transformation Groups of Polynomials over GF(2) with Applications to Linear Shift Register Sequences," *Information Sciences*, Vol. 1, 1968, pp. 87-109.

[119]   Golomb, S. W., "Some Decompositions of the Integers from 0 to p^n-1," *Am. Math. Monthly*, Vol. 79, No. 2, February, 1972, pp. 154-157.

[120]   Golomb, S. W., "Cyclotomic Polynomials and Factorization Theorems," *Am. Math. Monthly*, Vol. 85, No. 9, November, 1978, pp. 734-737.

[121]   Golomb, S. W., "Obtaining Specified Irreducible Polynomials Over Finite Fields," *SIAM J. Algebraic and Discrete Methods*, Vol. 1, No. 4, December, 1980.

[122]   Golomb, S. W., "On the Classification of Balanced Binary Sequences of Period 2^n-1," *IEEE Transactions on Information Theory*, vol. IT-26, 1980, pp. 730-732.

[123] Golomb, S.W., "Correlation Properties of Periodic and Aperiodic Sequences, and Applications to Multi-User Systems," in *Proceedings of the NATO Advanced Study Institute on Multi-User Communications* (Norwich, England, 1980), Sijthoff and Noordhoff, Alphen aan den Rijn, The Netherlands, 1981.

[124] Golomb, S. W., and L. D. Baumert, "Backtrack Programming," *Journal of Assoc. for Comp. Machinery*, Vol. 12, No. 4, October, 1965, pp. 516-524.

[125] Golomb, S.W., and L. D. Baumert, "The Search for Hadamard Matrices," *American Mathematical Monthly*, vol. 70, 1963, pp. 12-17.

[126] Golomb, S. W., and B. Gordon, "Codes with Bounded Synchronization Delay," *Information and Control*, Vol. 8, No. 4, August 1965, pp. 355-372.

[127] Golomb, S. W., B. Gordon, and L. R. Welch, "Comma-Free Codes," *Can. J. Math.*, Vol. 10, 1958, pp. 202-209.

[128] Golomb, S. W., and A. Lempel, "Second Order Polynomial Recursions," *SIAM J. Appl. Math.*, Vol. 33, No. 4, December, 1977, pp. 587-592.

[129] Golomb, S. W., and R. A. Scholtz, "Generalized Barker Sequences," *IEEE Transactions on Information Theory*, Vol. IT-11, 1965, pp. 533-537.

[130] Golomb, S. W., L. R. Welch, and M. Delbrück, "Construction and Properties of Comma-Free Codes," *Biol. Medd. Dan. Vid. Selsk.*, Vol. 23, No. 9, 1958.

[131] Golomb, S. W., L. R. Welch, and R. M. Goldstein, *Cycles from Nonlinear Shift Registers*, Prog. Report No. 20-389, JPL, California Institute of Technology, Pasadena, California, 1959.

*[132] Goncharov, V., "Du Domaine d'Analyse Combinatoire," *Bull. Acad. Sci. URSS, Ser. Math.*, Vol. 8, 1944, pp. 3-48.

*[133] Good, I. J., "Normal Recurring Decimals," *J. London Math. Soc.*, Vol. 21 (Part 3), 1946, pp. 167-172.

[134] Green, R. R., "A Serial Orthogonal Decoder," *JPL Space Program Summary 37-39*, Vol. 4, 1966, pp. 247-252.

[135] Griswold, R. E., *A Study of Iterative Switching Networks*, Technical Report No. 098-2, Stanford Electronics Laboratories, Stanford University, Stanford, California, 1962.

{136] Guibas, L. J., and A. M. Odlyzko, "Maximal Prefix-Synchronized Codes," *SIAM J. Appl. Math.*, Vol. 35, No. 2, September, 1978, pp. 401-418.

*[137] Hall, Marshall, Jr., "A Survey of Difference Sets," *Proc. Am. Math. Soc.*, Vol. 7, 1956, pp. 975-986.

[138] Hall, Marshall, Jr., *Combinatorial Theory*, Blaisdell Publishing Co., Waltham, Mass., 1967.

[139] Hardy, G. H., and E. M. Wright, *An Introduction to the Theory of Numbers*, 4th ed., Clarendon Press, Oxford, 1960.

[140] Harmuth, H. F., "Applications of Walsh Functions in Communications," *IEEE Spectrum*, Vol. 6, November, 1969, pp. 82-91.

*[141] Harrison, M. A., "On the Number of Classes of (n,k) Switching Networks," *J. Franklin Institute*, Vol. 276, No. 4, October, 1963, pp. 313-327.

[142] Harvey, J. T., "Delay Line in Shift Register Speeds m-Sequence Generation," *Electronics*, Vol. 48, No. 24, November, 1975, pp. 104-105.

[143] Harwit, M., and N. J. A. Sloane, *Hadamard Transform Optics*, Academic Press, New York, 1979.

[144] Healy, T. J., et al., *Statistical Properties of Weighted Random and Pseudo-Random Sequences*, Final Report for NASA Ames Research Center, N70-36008, University of Santa Clara, California, June, 1970.

[145] Hedlund, G. A., "Endomorphisms and Automorphisms of the Shift Dynamical System," *Math. Systems Theory*, Vol. 3, MR 41, #4510, 1969, pp. 320-375.

[146] Heimiller, R. C., "Phase Shift Pulse Codes with Good Periodic Correlation Properties," *IRE Transactions on Information Theory*, Vol. IT-7, 1961, pp. 254-257.

239

[147] Helleseth, T., "Some Results About the Cross-Correlation Function Between Two Maximal Linear Sequences," *Discrete Mathematics*, Vol. 16, 1976, pp. 209-232.

[148] Helleseth, T., "A Note on the Cross-Correlation Function Between Two Binary Maximal Length Linear Sequences," *Discrete Mathematics*, Vol. 23, 1978, pp. 301-307.

[149] Hemmati, F., and D. Costello, Jr., "An Algebraic Construction for Q-ary Shift Register Sequences," *IEEE Transactions on Computers*, Vol. C-27, No. 12, December, 1978, pp. 1192-1195.

[150] Henderson, K. W., "Comment on Computation of the Fast Walsh-Fourier Transform," *IEEE Transactions on Computers*, Vol. C-19, 1970, p. 850.

[151] Herlestam, T., "On Linearization of Nonlinear Combinations of Linear Shift Register Sequences," *1977 IEEE International Symposium on Information Theory*, Ithaca, New York, October, 1977.

[152] Hong, S. J., M. Y. Hsiao, E. I. Muehldorf, and A. M. Patel, "Linear Feedback Shift Register Sequence Generation for Specified Sequences," *IBM Tech. Disclosure Bull.*, Vol. 15, No. 8, January, 1973, pp. 2415-2416.

[153] Huffman, D. A., "The Synthesis of Linear Sequential Coding Networks," *Proc. of 3rd London Symposium on Information Theory*, Academic Press, New York, September, 1955, pp. 77-95.

[154] Huffman, D. A. "The Generation of Impulse-Equivalent Pulse Trains," *IRE Transactions on Information Theory*, Vol. IT-8, 1962, pp. S10-S16.

[155] Huffman, D. L., "A Modification of Huffman Impulse-Equivalent Pulse Trains to Increase Signal Energy Utilization," *IEEE Transactions on Information Theory*, Vol. IT-20, 1974, pp. 559-561.

[156] *IEEE Transactions on Communications: Special Issue on Spread Spectrum Communications*, Vol. COM-25, No. 8, 1977.

[157] Ikai, T., H. Kosako, Y. Kojima, "Ternary Pseudo-Random Noise Generator," *University of Osaka Prefecture Bul.*, Ser. A., Vol. 17, No. 1, 1968.

[158] Jordan, H. R., and D. C. M. Wood, "On the Distribution of Sums of Sucessive Bits of Shift-Register Sequences," *IEEE Transactions on Computers*, Vol. C-22, No. 4, April, 1973, pp. 400-408.

[159] Kabatyanskii, G. A., and V. I. Levenshtein, "Bounds for Packings on a Sphere and in Space," *Problemy Peredachi Informatsii*, Vol. 14, January, 1978, pp. 3-25, (in Russian), English translation in *Problems in Information Transmission*, Vol. 14, 1978, pp. 1-17.

[160] Kaiser, J., J. W. Schwartz, and J. M. Aein, "Multiple Access to a Communication Satellite with a Hard-Limiting Repeater. Volume I: Modulation Techniques and Their Applications," Inst. Defense Analyses, Rep. R-108, (AD 457945), 1965.

[161] Kasami, T., "Weight Distribution Formula for Some Class of Cyclic Codes," Coordinated Science Lab., University of Illinois, Urbana, Ill., Tech. Rep. R-285, (AD 632574), 1966.

[162] Kasami, T., "Weight Distribution of Bose-Chaudhuri-Hocquenghem Codes," *Combinatorial Mathematics and Its Applications*, Chapel Hill, NC: University of North Carolina Press, 1969. (Also Coordinated Science Lab., University of Illinois, Urbana, Ill., Tech. Rep. R-317, 1966).

[163] Kavanagh, R. J., "Fourier Analysis of Pseudo-Random Binary Sequences," *Electronics Letters*, Vol. 5, Part I, No. 7, April, 1969, pp. 173-174.

*[164] Kelly, J. B., "A Characteristic Property of Quadratic Residues," *Proc. Am. Math. Soc.*, Vol. 5, 1954, p. 38.

[165] Kendall, W. B., and I. S. Reed, "Path Invariant Comma-Free Codes," *IRE Transactions on Information Theory*, Vol. IT-8, October, 1962, pp. 350-355.

[166] Kerdock, A. M., F. J. MacWilliams, and A. M. Odlyzko, "A New Theorem About the Mattson-Solomon Polynomial and Some Applications," *IEEE Transactions on Information Theory*, Vol. IT-20, 1974, pp. 85-89.

[167] Key, E. L., "An Analysis of the Structure and Complexity of Nonlinear Binary Sequence Generators," *IEEE Transactions on Information Theory*, Vol. IT-22, 1976, pp. 732-736.

240

[168] Kjeldsen, K., "On the Cycle Structure of a Set of Nonlinear Shift Registers with Symmetric Feedback Functions," *J. Combinatorial Theory, Ser. A.*, Vol. 20, 1976, pp. 154-169.

[169] Kløve, T., "Linear Recurring Sequences in Boolean Rings," *Math. Scand.*, Vol. 33, 1973, pp. 5-12.

[170] Knuth, D. E., "Oriented Subtrees of an Arc Digraph," *J. Comb. Th.*, Vol. 3, No. 4, December, 1967, pp. 309-314.

[171] Knuth, Donald E., *The Art of Computer Programming, Vol., 2, Seminumerical Algorithms*, Addison-Wesley Publishing Co., Reading, Mass., 1969, p. 8, Exercise 13, and p. 454.

[172] Knuth, Donald E., *The Art of Computer Programming, Vol. 3: Sorting and Searching*, Addison-Wesley, 1973.

[173] Knuth, Donald E., and Luis Trabb-Pardo, "Analysis of a Simple Factorization Algorithm," *Theoretical Computer Science*, Vol. 3, No. 3, December, 1976, pp. 321-348.

[174] Kotov, Y. I., "Correlation Functions of Composite Sequences Constructed from m-Sequences," *Radio Eng. Electron. Phys.*, Vol. 19, 1974, pp. 128-130.

[175] Krishnaiyer, R., and J. C. Donovan, "Shift Register Generation of Pseudorandom Binary Sequences," *Comput. Des.*, Vol. 12, No. 4, April, 1973, pp. 69-74.

[176] Kriz, T. A., "Some Binary Output Sequence Properties of Deterministic Autonomous Finite-State Machines with Probabilistic Initialization," *IEEE Transactions on Computers*, Vol. C-22, No. 9, September, 1973.

[177] Kubista, T. T., et al., *Pseudo-Linear Generalized Shift-Registers*, Tech. Report EE-664, Department of Electrical Engineering, University of Notre Dame, Notre Dame, Indiana, June, 1966.

[178] Kummer, E. E., "Über die Divisoren gewisser Formen der Zahlen, welche aus der Theorie der Kreistheilung entstehen," *J. reine u. angew. Math.*, Vol. 30, 1846, pp. 107-116.

[179] Kummer, E. E., "Über die Zerlegung der aus Wurzeln der Einheit gebildeten complexen Zahlen in ihre Primfactoren," *J. reine u. angew. Math.*, Vol. 35, 1847, pp. 327-367.

[180] Laksov, D., "Linear Recurring Sequences Over Finite Fields," *Math. Scand.*, Vol. 16, 1965, pp. 181-196.

[181] Leach, E. B., "Regular Sequences and Frequency Distributions," *Proc. Amer. Math. Soc.*, Vol. 11, August, 1960.

[182] Lee, J., and D. R. Smith, "Families of Shift-Register Sequences with Impulsive Correlation Properties," *IEEE Transactions on Information Theory*, Vol. IT-20, 1974, pp. 255-261.

[183] Lempel, A., *Hadamard and m-Sequence Transforms are Permutationally Similar*, Applied Optics, Vol. 18 (24), pp. 4064-4065.

[184] Lempel, A., "On k-Stable Feedback Shift Registers," *IEEE Transactions on Computers*, Vol. 18, July, 1969, pp. 652-660.

[185] Lempel, A., "On a Homomorphism of the de Bruijn Graph and its Applications to the Design of Feedback Shift Registers," (IEEE Computer Group Repository Paper R-69-104, 1969), *IEEE Transactions on Computers*, Vol. C-19, No. 12, December, 1970, pp. 1204-1209.

[186] Lempel, A., "Analysis and Synthesis of Polynomials and Sequences over GF(2)," *IEEE Transactions on Information Theory*, Vol. IT-17, 1971, pp. 297-303.

[187] Lempel, A., "M-ary Closed Sequences," *J. Comb. Th., Series A.*, Vol. 10, No. 3, May, 1971, pp. 253-258.

[188] Lempel, A., "On the Extremal Factors of the de Bruijn Graph," *J. Comb. Th. (B)*, Vol. 11, No. 1, August, 1971, pp. 17-27.

[189] Lempel, A., "Matrix Factorization Over GF(2) and Trace-Orthogonal Bases of $GF(2^n)$," *SIAM J. Comput.*, No. 4, 1975, pp. 175-186.

[190] Lempel, A., and M. Cohn, *A Fast m-Sequence Transform*, Sperry Rand Research Center Report SRRC-RR-70-49, September, 1970.

[191] Lempel, A., M. Cohn, and W. L. Eastman, "A Class of Balanced Binary Sequences with Optimal Autocorrelation Properties," *IEEE Transactions on Information Theory*, Vol. IT-23, 1977, pp. 38-42.

[192] Lempel, A., and W. L. Eastman, "High Speed Generation of Maximal Length Sequences," *IEEE Transactions on Computers*, Vol. C-20, No. 2, February, 1971, pp. 227-229.

[193] Lempel, A., and H. Greenberger, "Families of Sequences with Optimal Hamming Correlation Properties," *IEEE Transactions on Information Theory*, Vol. IT-20, 1974, pp. 90-94.

[194] Lerner, R. M., "Signals Having Good Correlation Functions," *WESCON Conv. Rec.*, 1961.

[195] Levenshtein, V. I., "Binary Codes for the Correction of Deletions, Insertions, and Changes of Symbols," *Dokl. Acad. Nauk SSSR*, Vol. 163, August, 1965, pp. 845-848.

[196] Levitt, B. K., "Long Frame Sync Words for Binary PSK Telemetry," *IEEE Transactions on Communications*, Vol. COM-23, No. 11, November, 1975, pp. 1365-1367.

[197] Levitt, K. N., and J. K. Wolf, "On the Interleaving of Two-Level Periodic Binary Sequences," *Proc. N.E.C.*, 1965, pp. 644-649.

[198] Lindholm, J. H., "An Analysis of the Pseudo-Randomness Properties of Subsequences of Long m-Sequences," *IEEE Transactions on Information Theory*, Vol. IT-14, 1968, pp. 569-576.

[199] Lindner, J., "Binary Sequences Up to Length 40 with Best Possible Autocorrelation Function," *Electronics Letters*, Vol. 11, 1975, p. 507.

[200] Liu, C. L., and C. Tseng, "Complementary Sets of Sequences," *IEEE Transactions on Information Theory*, Vol. IT-18, September, 1972, pp. 644-651.

[201] Looft, F. J., *An Augmented Linear Maximal Shift Register Sequences Generator*, Michigan University, Ann Arbor, Cooley Electronics Lab., Office of Naval Research, Arlington, Va., September, 1975.

[202] Lorens, C. S., *The Generation of Nonlinear Sequences*, Report 1956, 3-089, Aerospace Corporation, June, 1964.

[203] MacWilliams, F. J., "An Example of Two Cyclically Orthogonal Sequences with Maximum Period," *IEEE Transactions on Information Theory*, Vol. IT-13, 1967, pp. 338-339.

[204] MacWilliams, F. J., and N. J. A. Sloane, "Pseudo-Random Sequences and Arrays," *Proceedings of the IEEE*, Vol. 64, 1976, pp. 1715-1729.

[205] MacWilliams, F. J., and N. J. A. Sloane, *The Theory of Error-Correcting Codes*, North-Holland, Amsterdam, the Netherlands, 1977.

[206] Magelby, K. R., *The Synthesis of NLFSR*, TR 6207-1, Stanford Electronic Lab., Stanford University, Stanford, Ca., October, 1963.

[207] Mandelbaum, D., "Arithmetic Codes with Large Distance," *IEEE Transactions on Information Theory*, Vol. 13, No. 2, April, 1967.

[208] Maritsas, D. G., "On the Statistical Properties of a Class of Linear Product Feedback Shift-Register Sequences," *IEEE Transactions on Computers*, Vol. C-22, No. 10, pp. 961-2.

*[209] Marsh, R. W., *Table of Irreducible Polynomials over GF(2) through Degree 19*, distributed by Office of Technical Services, Commerce Department, Washington, D. C., October, 1957.

**[210] Martin, M. H., "A Problem in Arrangements," *Bulletin of the American Mathematical Society*, Vol. 40, No. 12, December, 1934, pp. 859-864.

[211] Martin, R. L., *Studies of Feedback Shift-Register Synthesis of Sequential Machines*, The M.I.T. Press, 1969.

[212] Massey, J. L., "Some Algebraic and Distance Properties of Convolutional Codes," *Error Correcting Codes*, H. B. Mann, ed., John Wiley & Sons, New York, 1968, pp. 89-109.

[213] Massey, J. L., "Shift-Register Synthesis and BCH Decoding," *IEEE Transactions on Information Theory*, Vol. IT-15, January, 1969, pp. 122-127.

[214] Massey, J. L., "Optimum Frame Synchronization," *IEEE Transactions on Communications*, Vol. COM-20, No. 2, April, 1972, pp. 115-119.

[215] Massey, J. L., et al., "Monotone Feedback Shift Registers," *Proc. 2nd Allerton Conf.*, September, 1964.

[216] Massey, J. L., et al., "A New Diagram for Feedback Shift Registers," *Proc. 3rd. Allerton Conf.*, October, 1965, pp. 73-81.

[217] Massey, J. L., and J. J. Uhran, Jr., "Final Report for Multipath Study," Contract NAS 5-10786, University of Notre Dame, Notre Dame, Indiana, 1969.

[218] Massey, J. L., and J. J. Uhran, Jr., "Sub-baud Coding," *Proc. 13th Annual Allerton Conf. on Circuit and System Theory*, 1975, pp. 539-547.

[219] Mattson, H.F., and R. J. Turyn, *On Correlation by Subsequences*, Research Note No. 692, Sylvania ARL, February, 1967.

[220] Mattson, H. F., and G. Solomon, "A New Treatment of Bose-Chaudhuri Codes," *J. SIAM*, Vol. 9, 1961, pp. 654-669.

[221] Maury, J. L., and F. J. Styles, "Development of Optimum Frame Synchronization Codes for Goddard Space Flight Center PCM Telemetry Standards," *Proceedings of the 1965 National Telemetering Conference*, Los Angeles, California, June, 1964, Paper 3-1.

[222] McEliece, R. J., "On Periodic Sequences from GF(q)," *J. Combinatorial Theory, Series A*, Vol. 10, 1971, pp. 80-91.

[223] McEliece, R. J., "Correlation Properties of Sets of Sequences Derived from Irreducible Cyclic Codes," *Information and Control*, Vol. 45, 1980, pp. 18-25.

[224] McEliece, R. J., and H. Rumsey, Jr. "Euler Products, Cyclotomy, and Coding," *J. Number Theory*, Vol. 4, 1972, pp. 302-311.

[225] Mendelsohn, N. S., "Directed Graphs with the Unique Path Property," *Combinatorial Theory and Its Applications II*, Colloquia Mathematica Societatis Janos Bolyai, Edited by P. Erdös, A. Renyi, and Vera T. Sós, 1970, North Holland Publishing Co., pp. 783-799.

[226] Meshkovskii, K. A., "A New Class of Pseudorandom Sequences of Binary Signals," *Problemy Peredachi Informatsii*, Vol. 9, July, 1973, pp. 117-119. English translation in *Problems of Information Transmission*, Vol. 9, 1973, pp. 267-269.

[227] Mills, W. H., and Neal Zierler, "On a Conjecture of Golomb," *Pacific Journal of Mathematics*, Vol. 28, No. 3, March, 1969, pp. 635-640.

[228] Milstein, L. B., "Some Statistical Properties of Combination Sequences," *IEEE Transactions on Information Theory*, Vol. IT-23, 1977, pp. 254-258.

[229] Milstein, L. B. and R. R. Ragonetti, "Combination Sequences for Spread Spectrum Communications," *IEEE Transactions on Communications*, Vol. COM-25, 1977, pp. 691-696.

[230] Mitchell, William C., "An Evaluation of Golomb's Constant," *Mathematics of Computation*, Vol. 22, No. 102, April, 1968, pp. 411-415.

[231] Moon, J. W., and L. Moser, "On the Correlation Function of Random Binary Sequences," *SIAM J. Appl. Math.*, Vol. 16, No. 2, 1968, pp. 340-343.

[232] Moore, E. F., *Sequential Machines: Selected Papers*, Addison-Wesley, Reading, Mass., 1964.

[233] Mossige, S., "Constructive Theorems for the Truth Table of the Ford Sequence," *J. Comb. Th.(A)*, Vol. 11, No. 1, July, 1971, pp. 106-110.

[234] Mowle, F. J., *Enumeration and Classification of Stable Feedback Shift Registers*, Tech. Report EE-661, Department of Electrical Engineering, University of Notre Dame, Notre Dame, Indiana, January, 1966.

[235] Mowle, F. J., "Relations Between PN Cycles and Stable FSR," *IEEE Transactions on Electronic Computers*, Vol. EC-15, No. 3, June, 1966.

[236] Mowle, F. J., "An Algorithm for Generating Stable FSR of Order n," *J. Assoc. Comp. Machinery*, Vol. 14, No. 3, July, 1967.

[237] Mykkeltveit, J., "A Proof of Golomb's Conjecture for the de Bruijn Graph," *Journal of Combinatorial Theory (B)*, Vol. 13, No. 1, August, 1972, pp. 40-45.

[238] Mykkeltveit, J., "Generating and Counting the Double Adjacencies in a Pure Circulating Shift Register," *IEEE Transactions on Computers*, Vol. C-24, No. 3, March, 1975.

[239] Mykkeltveit, J., *Generalization of a Theorem on Linear Recurrences to the Non-Linear Case*. Internal Report, University of Bergen, Bergen, Norway, 1976.

[240] Mykkeltveit, J., "Nonlinear Recurrences and Arithmetic Codes," *Information and Control*, Vol. 33, No. 3, March, 1977.

[241] Mykkeltveit, J., and E. S. Selmer, "Linear Recurrence in Boolean Rings. Proof of Kløve's Conjecture," *Math. Scand.*, Vol. 33, 1973, pp. 13-17.

¶[242] Nadler, M., and A. Sengupta, "Shift Register Code for Indexing Applications," *Communications of the ACM*, Vol. 2, No. 10, October, 1959, pp. 40-43.

[243] Nathanson, M. B., "Derivatives of Binary Sequences," *SIAM J. on Appl. Math.*, Vol. 21, No. 3, November, 1971, pp. 407-412.

[244] Nelson, E. D., and M. L. Fredman, "Hadamard Spectroscopy," *J. Opt. Soc. Amer.*, Vol. 60, 1970, pp. 1664-1669.

[245] Neuman, F., and L. Hofman, "New Pulse Sequences with Desirable Correlation Properties," *1971 National Telemetering Conference Record*, Washington, D. C., pp. 277-282.

[246] Nielsen, P. T., "Distributed Frame Synchronization by Nonlinear-Shift-Register Sequences," *Electronics Letters*, Vol. 9, No. 4, February, 1973, pp. 73-75.

[247] Nielsen, P. T., "Some Optimum and Suboptimum Frame Synchronizers for Binary Data in Gaussian Noise," *IEEE Transactions on Communications*, Vol. COM-21, No. 6, June, 1973, pp. 770-772.

[248] Nielsen, P. T., "On the Expected Duration of a Search for a Fixed Pattern in Random Data," *IEEE Transactions on Information Theory*, Vol. IT-19, No. 5, September, 1973, pp. 702-704.

→[249] Niho, Y. "Multi-Valued Cross-Correlation Functions Between Two Maximal Linear Recursive Sequences," Ph.D. Dissertation, Department of Electrical Engineering, University of Southern California, Los Angeles, CA (also USC EE Rep. 409), 1972.

[250] Nikforuk, P. N., and M. M. Gupta, "A Bibliography on the Properties, Generation and Control System Applications of Shift-Register Sequences," *Int. Jnl. Control*, Vol. 9, No. 2, 1969, pp. 217-234.

[251] Norton, K. K., "Numbers with Small Prime Factors, and the Least k^{th} Power Non-Residue," *Memoirs Amer. Math. Soc.*, Vol. 106, 1971, pp. 9-27.

[252] Olsen, J. D., *Non-Linear Binary Sequences with Asymptotically Optimum Periodic Cross-Correlation*, Dissertation, University of Southern California, Los Angeles, CA, December, 1977.

[253] Olsen, J. D., and C. M. Heard, "A Comparison of the Visual Effects of Two Transform Domain Encoding Approaches," *Society of Photo-Optical Instrumentation Engineers Semin. Proc.*, Vol. 119, Bellingham, Washington, 1977, pp. 137-146.

[254] Olsen, J.D., R.A. Scholtz, and L.R. Welch, "Bent Function Sequences," to appear in *IEEE Transactions on Information Theory*, November, 1982.

*[255] Ore, Ø., "Contributions to the Theory of Finite Fields," *Transactions of the Am. Math. Soc.*, Vol. 36, 1934, pp. 243-274.

[256] Paaske, E., et al., "Note on Incoherent Binary Sequences," *IEEE Transactions on Aerospace and Electronic Systems*, Vol. AES-4, No. 1, January, 1968.

*[257] Paley, R. E. A. C., "On Orthogonal Matrices," *J. Math. Phys.*, Vol. 12, 1933, pp. 311-320.

[258] Payne, W. H., and K. L. McMillen, "Orderly Enumeration of Nonsingular Binary Matrices Applied to Text Encryption," *Commun. ACM.*, Vol. 21, No. 4, April, 1978, pp. 259-263.

[259] Pelekhatyi, M. I., and E. A. Golubev, "Autocorrelative Properties of Certain Types of Binary Sequences," *Problemy Peredachi Informatsii*, Vol. 8, January, 1972, pp. 92-99, (in Russian). English translation in *Problems of Information Transmission*, Vol. 8, 1972, pp. 71-76.

[260] Perlman, M., "Decomposition of States of a Linear Feedback Shift Register into Cycles of Equal Length," *JPL Space Programs Summary 37-52*, Vol. 3, August, 1968, pp. 149-154.

¶[261] Perron, O., "Bemerkungen über die Verteilung der Quadratischen Reste," *Math. Zeitschrift*, Vol. 56, 1952, pp. 122-130.

¶[262] Peterson, W. W., *Error Correcting Codes*, John Wiley and Sons, Inc., New York, 1961.

[263] Peterson, W. W., and E. J. Weldon, Jr., *Error-Correcting Codes*, 2nd Ed., M.I.T. Press, Cambridge, Mass., 1972.

[264] Pohlig, S. C., and M. E. Hellman, "An Improved Algoithm for Computing Logarithms Over GF(p) and Its Cryptographic Significance," *IEEE Transactions on Information Theory*, Vol. IT-24, No. 1, January, 1978, pp. 106-110.

[265] Pollard, J. M., "A Monte Carlo Method for Factorization," *BIT*, Vol. 15, 1975, pp. 331-334.

[266] Posner, E. C., "Combinatorial Structures in Planetary Reconnaissance," *Error Correcting Codes*, H. B. Mann, ed., Wiley, New York, 1968, pp. 15-46.

[267] Purdom, Paul W., and John H. Williams, "Cycle Length in a Random Function," *Transactions of the American Mathematical Society*, Vol. 133, No. 2, September, 1968, pp. 547-551.

[268] Pursley, M. B. "Evaluating Performance of Codes for Spread Spectrum Multiple Access Communications," *Proceedings of the 12th Annual Allerton Conference on Circuit and System Theory*, 1974, pp. 765-774.

[269] Pursley, M. B., "Performance Evaluation for Phase-Coded Spread-Spectrum Multiple-Access Communication-Part I: System Analysis," *IEEE Transactions on Communications*, Vol. COM-25, 1977, pp. 795-799.

[270] Pursley, M. B., "On the Mean-Square Partial Correlation of Periodic Sequences," *Proceedings of the 1979 Conference on Information Sciences and Systems*, Johns Hopkins University, Baltimore, Md., 1979, pp. 377-379.

[271] Pursley, M. B., and F. D. Garber, "Quadriphase Spread-Spectrum Multiple-Access Communications," *IEEE Int. Conf. on Communications, Conf. Record.*, Vol. 1, 1978, pp. 7.3.1.-7.3.5.

[272] Pursley, M. B., and H. F. A. Roefs, "Numerical Evaluation of Correlation Parameters for Optimal Phases of Binary Shift-Register Sequences," *IEEE Transactions on Communications*, Vol. COM-27, 1979, pp. 1597-1604.

[273] Pursley, M. B., and D. V. Sarwate, "Bounds on Aperiodic Cross-Correlation for Binary Sequences," *Electronics Letters*, Vol. 12, 1976, pp. 304-305.

[274] Pursley, M. B., and D. V. Sarwate, "Evaluation of Correlation Parameters for Periodic Sequences," *IEEE Transactions on Information Theory*, Vol. IT-23, 1977, pp. 508-513.

[275] Pursley, M. B., and D. V. Sarwate, "Performance Evaluation for Phase-Coded Spread-Spectrum Multiple-Access Communication — Part II: Code Sequence Analysis," *IEEE Transactions on Communications*, Vol. COM-25, 1977, pp. 800-803.

[276] Ramaswami, V., "The Number of Positive Integers $< x$ and Free of Prime Divisor $> x^c$, and a Problem of S. S. Pillai," *Duke Math. J.*, Vol. 16, 1949, pp. 99-109.

[277] Reed, I. S., and H. Blasbalg, "Multipath Tolerant Ranging and Data Transfer Techniques for Air-to-Ground and Ground-to-Air Links," *Proceedings of the IEEE*, Vol. 58, 1970, pp. 422-429.

[278] Reed, I. S., and T. K. Truong, "The Use of Finite Fields to Compute Convolutions," *IEEE Transactions on Information Theory*, Vol. IT-21, 1975, pp. 208-213.

[279] Reed, I. S., and R. Turn, "A Generalization of Shift-Register Sequence Generators," *J. Assoc. Comput. Mach.*, Vol. 16, No. 3, July, 1969, pp. 461-473.

[280] Rees, D., "Notes on a Paper by I. J. Good," *J. London Math. Soc.*, Vol. 21, (Part 3), 1946, pp. 169-172.

[281] Reiffen, B., and H.L. Yudkin, "On Nonlinear Binary Sequential Circuits and Their Inverses," *IEEE Transactions on Electronic Computers*, vol. EC-15, no. 4, August, 1966, pp. 586-596.

[282] Reuschle, K. G., *Tafeln Complexer Primzahlen, welche aus Wurzeln der Einheit gebildet sind: Auf dem Grunde Kummerschen Theorie der complexen Zahlen*, Akademie der Wissenschaften, Berlin, 1875.

[283] Riordan, J., *An Introduction to Combinatorial Analysis*, John Wiley & Sons, Inc., New York, 1958.

[284] de Rivière, A., "Question No. 48," *L'Intermédiare des Mathématiciens*, Vol. 1, No. 2, February, 1894, pp. 19-20.

[285] Roefs, H. F. A., "Binary Sequences for Spread-Spectrum Multiple-Access Communication," Ph.D. Dissertation, Department of Electrical Engineering, University of Illinois, Urbana, Ill., (also Coordinated Science Lab. Report R-785), August, 1977.

[286] Roefs, H. F. A., and M. B. Pursley, "Correlation Parameters of Random Binary Sequences," *Electronics Letters*, Vol. 13, 1977, pp. 488-489.

[287] Roefs, H. F. A., D. V. Sarwate, and M. B. Pursley, "Periodic Correlation Functions for Sums of Pairs of m-Sequences," *Proc. 1977 Conf. Information Sciences and Systems*, Johns Hopkins University, Baltimore, Md., 1977, pp. 487-492.

[288] Roth, H. H., "Linear Binary Shift Register Circuits Utilizing a Minimum Number of Mod-2 Adders," *IEEE Transactions on Information Theory*, Vol. IT-11, April, 1965, pp. 215-220.

**[289] Rothaus, O. S., "On Bent Functions," *Journal of Combinatorial Theory (A)*, Vol. 20, No. 3, May, 1976, pp. 300-305.

*[290] Ryser, H. J., *Combinatorial Mathematics*, The Carus Mathematical Monographs, Mathematical Association of America, John Wiley & Sons, Inc., 1963.

** [291] Sainte-Marie, C. Flye, "Solution to Question No. 48", *L'Intermédiare des Mathématiciens*, Vol. 1, No. 6, June, 1894, pp. 107-110.

[292] Sarwate, D. V., "Comments on 'A Class of Balanced Binary Sequences with Optimal Autocorrelation Properties'," *IEEE Transactions on Information Theory*, Vol. IT-24, 1978, pp. 128-129.

[293] Sarwate, D. V., "Cross-correlation Properties of Sequences with Applications to Spread-Spectrum Multiple-Access Communication," *Proc. AFOSR Workshop in Communication Theory and Applications*, Provincetown, Mass., 1978, pp. 88-91.

[294] Sarwate, D. V., "Bounds on Cross-correlation and Autocorrelation of Sequences," *IEEE Transactions on Information Theory*, Vol. IT-25, 1979, pp. 720-724.

[295] Sarwate, D. V., and M. B. Pursley, "Applications of Coding Theory to Spread-Spectrum Multiple-Access Satellite Communications," *Proc. 1976 IEEE Canadian Communications and Power Conf.*, 1976, pp. 72-75.

[296] Sarwate, D. V., and M. B. Pursley, "New Correlation Identities for Periodic Sequences," *Electronic Letters*, Vol. 13, No. 2, 1977, pp. 48-49.

[297] Sarwate, D. V., and M. B. Pursley, "Hopping Patterns for Frequency Hopped Multiple-Access Communications," *IEEE Int. Conf. Communications, Conf. Rec.*, 1978, pp. 7.4.1.-7.4.3.

**[298] Sarwate, D. V., and M. B. Pursley, "Cross-correlation Properties of Pseudorandom and Related Sequences," *Proceedings of the IEEE*, Vol. 68, No. 5, May, 1980, pp. 593-619.

[299] Sastri, K. S., "Specified Sequence Linear Feedback Shift Registers," *Int. J. Syst. Sci.*, Vol. 6, No. 11, November, 1975, pp. 1009-1019.

[300] Schneider, K. S., and R. S. Orr, "Aperiodic Correlation Constraints on Large Binary Sequence Sets," *IEEE Transactions on Information Theory*, Vol. IT-21, 1975, pp. 79-84.

[301] Scholtz, R. A., "Automated Code Word Synchronization — The Noiseless Case," *Proc. of the Inter. Conf. on Communications*, San Francisco, Calif., June, 1970, pp. 18.8-18.14.

[302] Scholtz, R. A., *Optimal CDMA Codes*, NTC Conference Record, November, 1979.

[303] Scholtz, R. A., "Codes with Synchronization Capability," *IEEE Transactions on Information Theory*, Vol. IT-12, April, 1966, pp. 135-142.

[304] Scholtz, R. A., "Maximal and Variable Word-Length Comma-Free Codes," *IEEE Transactions on Information Theory*, Vol. IT-15, March, 1969, pp. 300-306.

[305] Scholtz, R. A., "The Search for Digital Radar-Ranging Signals," *Proceedings of the 4th Hawaii International Conference on System Sciences*, January, 1971, pp. 377-379.

[306] Scholtz, R. A., and L. R. Welch, "Generalized Residue Sequences," *ICC Conference Record*, June, 1973.

[307] Scholtz, R. A., and L. R. Welch, The Mechanization of Codes with Bounded Synchronization Delays," *IEEE Transactions on Information Theory*, Vol IT-16, July, 1970, pp. 438-446.

**[308] Scholtz, R. A., and L. R. Welch, "Group Characters: Sequences with Good Correlation Properties," *IEEE Transactions on Information Theory*, Vol. IT-24, No. 5, September, 1978, pp. 537-545.

[309] Schroeder, M. R., "Synthesis of Low-Peak-Factor Signals and Binary Sequences with Low Autocorrelation," *IEEE Transactions on Information Theory*, Vol. IT-16, 1966, pp. 79-84.

[310] Schwartz, J. W., J. M. Aein, and J. Kaiser, "Modulation Techniques for Multiple-Access to a Hard-Limiting Satellite Repeater," *Proceedings of the IEEE*, Vol. 54, 1966, pp. 763-777.

[311] Schweitzer, B. P., *Generalized Complementary Code Sets*, Ph.D. Dissertation, University of California, Los Angeles, 1971.

[312] Seguin, G., "Binary Sequences with a Relaxed Barker Criterion," *Proc. National Electronics Conference*, Vol. 26, December, 1970, pp. 456-457.

[313] Seguin, G., "Binary Sequences with Specified Correlation Properties," Ph.D. Dissertation, Department of Electrical Engineering, University of Notre Dame, Notre Dame, Indiana, (also Tech. Rep. 7103), 1971.

[314] Sellers, F. F., Jr., "Bit Loss and Gain Correction Code," *IRE Transactions on Information Theory*, January, 1962, pp. 35-38.

*[315] Selmer, E., "On the Irreducibility of Certain Trinomials," *Math. Scand.*, Vol. 4, 1956, pp. 287-302.

[316] Selmer, E., "Linear Recurrence Relations Over Finite Fields," Depart. Math., University of Bergen, Bergen, Norway, 1966.

[317] Seroussi, G., and A. Lempel, "Factorization of Symmetric Matrices and Trace-Orthogonal Bases in Finite Fields," *SIAM J. Comput.*, Vol. 9, No. 4, 1980, pp. 758-767.

*[318] Shannon, C. E., "Communication Theory of Secrecy Systems," *Bell System Tech. J.*, Vol. 28, No. 4, October, 1949, pp. 656-715.

[319] Shedd, D. A., and D. V. Sarwate, "Construction of Sequences with Good Correlation Properties," *IEEE Transactions on Information Theory*, Vol. IT-25, 1979, pp. 94-97.

[320] Shepp, L., and S. P. Lloyd, "Ordered Cycle Lengths in a Random Permutation," *Trans. Amer. Math. Soc.*, Vol. 121, 1966, pp. 340-357.

[321] Sidel'nikov, V. M., "Some k-Valued Pseudo-Random Sequences and Nearly Equidistant Codes," *Problemy Peredachi Informatsii*, Vol. 5, January, 1969, pp. 16-22 (in Russian). English translation in *Problems in Information Transmission*, Vol. 5, 1969, pp. 12-16.

[322] Sidel'nikov, V. M., "Cross-Correlation of Sequences," *Problemy Kybernetika*, 1971, pp. 15-42.

[323] Sidel'nikov, V. M., "On Mutual Correlation Sequences," *Soviet Math. Dokl.*, Vol. 12, 1971, pp. 197-201.

[324] Sinna, S., and P. E. Caines, "On the Use of Shift Register Sequences as Instrumental Variables for the Recursive Identification of Multivariable Linear Systems," *Int. J. Syst. Sci.*, Vol. 8, No. 9, September, 1977, pp. 1041-1055.

*[325] Slepian, D., "On the Number of Symmetry Types of Boolean Functions of n Variables," *Can. J. Math.*, Vol. 5, No. 2, pp. 185-193.

[326] Sloan, K. W., et al., *The Structure of Irreducible Polynomials Mod 2 Under a Cubic Transformation*, NSA Section 314 Report, July, 1953.

**[327] Slowinski, David, "Searching for the Twenty-Seventh Mersenne Prime," *Journal of Recreational Mathematics*, Vol. 11, No. 4, 1978-79, pp. 258-261.

[328] Smirnov, N. I., "Applications of m-Sequences in Asynchronous Radio Systems," *Telecommun. Radio Eng.*, Vol. 24, No. 10, 1970, pp. 26-35, (translated from the Russian journal Elektrosvyaz).

[329] Smirnov, N. I., and N. A. Golubkov, "Correlation Properties of Segments of m-Sequences," *Telecommun. Radio Eng.*, Vol. 28, No. 6, 1973, pp. 123-125, (translation from the Russian journal Elektrosvyaz).

[330] Smith, A. R., "General Shift-Register Sequences of Arbitrary Cycle Length," *IEEE Transactions on Computers*, Vol. C-20, No. 4, April, 1971, pp. 456-459.

[331] Solomon, G., "Optimal Frequency Hopping Sequences for Multiple-Access," *Proc. 1973 Symp. Spread Spectrum Communications*, Vol. 1 (AD 915852), 1973, pp. 33-35.

[332] Solomon, G., and R. J. McEliece, "Weights of Cyclic Codes," *J. Comb. Th.*, Vol. 1, 1966, pp. 459-475.

[333] Somaini, U., and M. H. Ackroyd, "Uniform Complex Codes with Low Autocorrelation Sidelobes," *IEEE Transactions on Information Theory*, Vol. IT-20, 1974, pp. 689-691.

[334] Søreng, J., "The Periods of the Sequences Generated by Some Symmetric Shift Registers," *J. Comb. Th.*, Ser. A., Vol. 21, 1976, pp. 164-187.

247

[335] Søreng, J., "Symmetric Shift Registers," *Pacific Journal of Math.*, vol. 85, no. 1, 1979, pp. 201-229.

[336] Søreng, J., "Symmetric Shift Registers, Part 2," *Pacific Journal of Math.*, vol. 98, no. 1, 1982, pp. 203-234.

[337] Spann, R., "A Two-Dimensional Correlation Property of Pseudo-Random Maximal Length Sequences," *Proceedings of the IEEE*, Vol. 53, Part 2, 1965, p. 2137.

[338] Spilker, J. J., Jr., "GPS Signal Structure and Performance Characteristics," *Navigation*, Vol. 25, No. 2, Summer, 1978, pp. 121-146.

[339] Spindel, R. C., K. R. Peal, and D. E. Koelsch, "A Microprocessor Acoustic Data Buoy," *The Ocean Challenge*, IEEE, New York, September, 1978, pp. 527-531.

[340] Stalder, J. E., and C. R. Cahn, "Bounds for Correlation Peaks of Periodic Digital Sequences," *Proceedings of the IEEE*, Vol. 52, 1964, pp. 1262-1263.

[341] Stiffler, J. J., "Rapid Acquisition Sequences," *IEEE Transactions on Information Theory*, Vol. IT-14, 1968, pp. 221-225.

[342] Stiffler, J. J., *Theory of Synchronous Communications*, Prentice-Hall, Englewood Cliffs, N.J., 1971.

[343] Stiglitz, I. G., "Multiple-Access Considerations — A Satellite Example," *IEEE Transactions on Communications*, Vol. COM-21, 1973, pp. 577-582.

[344] Storer, Thomas, *Cyclotomy and Difference Sets*, Markham Publishing Co., 1967.

[345] Su, *Shift Register Realizations of Sequential Machines*, Report PR-67-104, Systems Laboratory, Northwestern University, Evanston, Illinois, 1967, (AD 646593).

[346] Sugiyama, Y., S. Hirasawa, M. Kasahara, and T. Namekawa, "The Construction of Sequences by Interleaving Method," *Electronics and Communications in Japan*, Vol. 55-A, 1972, pp. 35-42.

[347] Surbock, F., and H. Weinrichter, "Interlacing Properties of Shift-Register Sequences with Generator Polynomials Irreducible Over GF(p)," *IEEE Transactions on Information Theory*, Vol. IT-24, No. 3, May, 1978, pp. 386-389.

[348] Susskind, A. K., et al., Report RADC-TDR-64-492, MIT, Cambridge, Mass., 1964.

[349] Susskind, A. K., "Realization of Sequential Machines in Feedback Shift Register Form," in *Network and Switching Theory*, G. Biorci (Ed.), Academic Press, 1968, pp. 558-594.

*[350] Swan, R. G., "Factorization of Polynomials Over Finite Fields," *Pacific J. Math.*, Vol. 12, 1962, pp. 1099-1106.

[351] Taki, Y., H. Miyakawa, M. Hatori, and S. Namba, "Even-Shift Orthogonal Sequences," *IEEE Transaction on Information Theory*, Vol. IT-15, 1969, pp. 295-300.

[352] Tan, C. J., *Factorization of Linear Cycle Sets*, Semiannual Report, Electronics Research Laboratory, University of California, Berkeley, Calif., 1965, pp. 230-232.

[353] Taub, H., and D. L. Schilling, *Principles of Communication Systems*, McGraw Hill, 1971.

*[354] Taussky, O., and J. Todd, "Generation of Pseudo-Random Numbers," *Symposium on Monte Carlo Methods*, H. A. Meyer, ed., John Wiley & Sons, Inc., New York, 1956, pp. 15-18.

[355] Titsworth, R. C., "Optimal Ranging Codes," *IEEE Transactions on Space Electronics and Telemetry*, Vol. SET-10, 1964, pp. 19-30.

[356] Tolstrup Nielsen, P., "Distributed Frame Synchronization by Nonlinear-Shift-Register Sequences," *Electronics Letters*, Vol. 9, No. 4, 1973, pp. 73-75.

[357] Tomlinson, G. H., and P. Galvin, "Properties of the Subsequences of a Class of Linear Product Sequences," *Proc. Institution of Electr. Eng.*, Vol. 122, No. 10, October, 1975, pp. 1095-1096.

[358] Tootill, J. P. R., W. D. Robinson, and D. J. Eagle, "An Asymptotically Random Tausworthe Sequence," *J. Assoc. Comput. Mach.*, Vol. 20, No. 3, July, 1973, pp. 469-481.

[359] Tseng, C. C., and C. L. Liu, "Complementary Sets of Sequences," *IEEE Transactions on Information Theory*, Vol. IT-18, 1972, pp. 644-652.

[360] Turyn, R., "The Correlation Function of a Sequence of Roots of 1," *IEEE Transactions on Information Theory*, Vol. IT-13, 1967, pp. 524-525.

[361] Turyn, R., "Sequences with Small Correlation," *Error Correcting Codes*, H. B. Mann, ed., Wiley, New York, 1968.

[362] Turyn, R., "Four-Phase Barker Codes," *IEEE Transactions on Information Theory*, Vol. IT-20, 1974, pp. 366-371.

[363] Turyn, R., and J. Storer, "On Binary Sequences," *Proc. Amer. Math. Soc.*, Vol. 12, 1961, pp. 394-399.

[364] Ullman, J. D., "Near Optimal, Single-Synchronization-Error Correcting Codes," *IEEE Transactions on Information Theory*, Vol. IT-12, No. 4, October, 1966, pp. 418-424.

[365] Ullman, J. D., "On the Capabilities of Codes to Correct Synchronization Errors," *IEEE Transactions on Information Theory*, Vol. IT-13, No. 1, June, 1967, pp. 95-105.

¶[366] van der Waerden, B. L., *Modern Algebra*, Vol. 1 (1949) and Vol. 2 (1950), Frederick Ungar Publishing Company, New York

[367] van Lantschoot, E. J. M., *Nonlinear Feedback Shift Registers: Analysis and Synthesis*, Ph.D. Dissertation, Catholic University of Louvain, Heverlee, Belgium, 1970.

[368] van Lint, J. H., *Coding Theory*, Springer, New York, 1971.

[369] van Rongen, J. B., "On the Largest Prime Divisor of an Integer," *Proc. Kon. Ned. Akad. Wet.*, Vol. A 78, 1975, pp. 70-76.

[370] Wainberg, S., and J. K. Wolf, "Subsequences of Pseudorandom Sequences," *IEEE Transactions on Communications*, Vol. COM-18, 1970, pp. 606-612.

[371] Wallis, W. D., A. D. Street, and J. S. Wallis, "Lecture Notes in Mathematics-Combinatorics: Room Squares, Sum-Free Sets, Hadamard Matrices," Springer-Verlag, 1972.

[372] Wang, K. C., *Synthesis of Linear Sequential Machines*, IEEE Computer Group Repository Paper R-67-8, 1967.

[373] Wang, K. C., "Algebraic Relations of Feedback Shift Register Sequences," *Proceedings of the 6th Hawaii International Conference on Systems Sciences, Supplement II*, Honolulu, Hawaii, January, 1973, pp. 66-69.

[374] Ward, M., "The Arithmetical Theory of Linear Recurring Sequences," *Trans. Amer. Math. Soc.*, Vol. 35, 1933, pp. 600-628.

[375] Weathers, G. D., E. R. Graf, and G. R. Wallace, "The Subsequence Weight Distribution of Summed Maximal Length Digital Sequences," *IEEE Transactions on Communications*, Vol. COM-22, 1974, pp. 997-1004.

➡[376] Welch, L. R., "Lower Bounds on the Maximum Cross-Correlation of Signals," *IEEE Transactions on Information Theory*, Vol. IT-20, 1974, pp. 397-399.

[377] Welti, G. R., "Quaternary Codes for Pulsed Radar," *IRE Transactions on Information Theory*, Vol. IT-6, June, 1960, pp. 400-405.

[378] Weng, L., "Decomposition of m-Sequences and Its Applications," *IEEE Transactions on Information Theory*, Vol. IT-17, 1971, pp. 457-463.

[379] Willard, M. W., "Optimum Code Patterns for PCM Synchronization," *Proceedings of the 1962 National Telemetering Conference*, Washington, D.C,, May, 1962, Paper 5-5.

[380] Willett, M., "The Minimum Polynomial for a Given Solution of a Linear Recursion," *Duke Math. Journal*, vol. 39, 1972, pp. 101-104.

[381] Willett, M., "Matrix Fields Over GF (q)," *Duke Math. Journal*, vol. 40, 1973, pp. 701-704.

[382] Willett, M., "Cycle Representatives for Minimal Cyclic Codes," *IEEE Transactions on Information Theory*, vol. IT-21, 1975, pp. 716-718.

[383] Willett, M., "On a Theorom of Kronecker," *The Fibonacci Quarterly*, vol. 14, 1976, pp. 27-29.

[384] Willett, M., "The Index of an m-Sequence," *SIAM J. Appl. Math.*, Vol. 25, 1973, pp. 24-27.

[385] Willett, M., "Characteristic m-Sequences," *Math. Comput.*, Vol. 30, 1976, pp. 306-311.

[386] Willett, M., "Factoring Polynomials Over A Finite Field," *SIAM J. Appl. Math.*, vol. 35, 1978, pp. 333-337.

[387] Willett, M., "Arithmetic in a Finite Field," *Mathematics of Computation*, vol. 35, pp. 1353-1358.

[388] Wu, W. W., "Applications of Error-Coding Techniques to Satellite Communications," *COMSAT Tech. Rev.*, Vol. 1, No. 1, Fall, 1971, pp. 183-219.

[389] Wu, W. W., *On Generalization of Generalized Shift Register Sequence Generators*, Technical Memo CL-37-70, COMSAT Labs, pp. 1-9, 1970.

[390] Wunderlich, M. L., and J. L. Selfridge, "A Design for a Number Theory Package with an Optimized Trial Division Routine," *Comm. ACM*, Vol. 17, 1974, pp. 272-276.

[391] Yao, K., "Error Probability of Asynchronous Spread Spectrum Multiple Access Communications Systems," *IEEE Transactions on Communications*, Vol. COM-25, 1977, pp. 803-809.

[392] Yoeli, M., "Nonlinear Feedback Shift Registers," IBM Development Lab., Poughkeepsie, N. Y., Tech. Report TR00.809, September, 1961.

[393] Yoeli, M., "Binary Ring Sequences," *Am. Math. Monthly*, Vol. 69, November, 1962, pp. 852-855.

[394] Yoeli, M., "Counting with Nonlinear Binary Feedback Shift Registers," *IEEE Transactions on Electronic Computers*, Vol. EC-12, August, 1962, pp. 357-361.

[395] Zetterberg, Lars H., "Communications Research in Sweden," *IEEE Transactions on Communications*, Vol. COM-22, No. 9, September 1974, pp. 1493-1498.

[396] Zierler, Neal, *Several Binary Sequence Generators*, TR 95, Lincoln Lab., MIT, September, 1955.

[397] Zierler, Neal, "On a Class of Binary Sequences," *Proc. Amer. Math. Soc.*, Vol. 7, 1956, pp. 675-681.

*[398] Zierler, Neal, "On the Theorem of Gleason and Marsh," *Proc. Amer. Math. Soc.*, Vol. 9, 1958, pp. 236-237.

[399] Zierler, Neal, "Linear Recurring Sequences," *J. Soc. Industrial and Applied Mathematics*, Vol. 7, 1959, pp. 31-48.

[400] Zierler, Neal, "Linear Recurring Sequences and Error-Correcting Codes," *Error Correcting Codes*, H. B. Mann, Ed., Wiley, New York, 1968.

[401] Zierler, Neal, "On $x^n + x + 1$ over GF(2)," *Information and Control*, Vol. 16, 1970, pp. 502-505.

[402] Zierler, Neal, "A Conversion Algorithm for Logarithms on $GF(2^n)$," *Journal of Pure and Applied Algebra*, Vol. 4, No. 3, June, 1974, pp. 353-356.

**[403] Zierler, Neal, and John Brillhart, "On Primitive Trinomials (mod 2)," *Information and Control*, Vol. 13, No. 6, December, 1968, pp. 541-554.

**[404] Zierler, Neal, and John Brillhart, "On Primitive Trinomials (mod 2) II," *Information and Control*, Vol. 14, No. 6, June, 1969, pp. 556-569.

[405] Zierler, N., and W. H. Mills, "Products of Linear Recurring Sequences," *J. Algebra*, Vol. 27, 1973, pp. 147-157.

[406] Zierler, N., "Linear Recurring Sequences," *Linear Sequential Switching Circuits*, W. Kautz, ed., Holden-Day, Inc., San Francisco, 1965.

*Reference is contained in References to Chapters I-VIII, page 223.
**Reference is contained in References to Chapter IX, page 229.